*L*AURENS VAN VELDHUIZEN was educated at the Agricultural University of Wageningen in the Netherlands, where he specialised in Extension Education and Soil and Water Management. From 1979 to 1983 he was engaged in research and development of the use of wind power for small-scale irrigation both overseas and at the International Institute for Land Reclamation and Improvement in the Netherlands. He then worked as extension advisor for the Reformed Church of the Netherlands in its Integrated Rural Development Programme in Eastern Indonesia. Since 1990, he has been a senior consultant with the ETC Foundation, supporting the development of operational approaches for participatory research and extension and the development of participatory training approaches and materials. His work has taken him to many parts of the world, most frequently to Asia. He is co-editor of *Linking with Farmers: Networking for Low-External-Input and Sustainable Agriculture* and *Farmers' Research in Practice*.

*A*NN WATERS-BAYER completed her doctorate in Rural Sociology at the University of Stuttgart-Hohenheim, after previous studies at the University of Winnipeg, Queen's University at Kingston, Canada and the University of New England in Armidale, Australia. From 1981 to 1985, she was involved as a socio-economist with the International Livestock Centre for Africa (ICLA) in collaborative research with Fulani agro-pastoralists in Nigeria. She then joined the German Institute for Tropical and Sub-tropical Agriculture in Witzenhausen. Since 1989, she has been a senior consultant with the ETC Foundation in Leusden, the Netherlands, where she advises research and development agents working with livestock-keepers in ecologically oriented technology development and natural resource management in various countries of Africa. This work has included six years on the staff of the ETC Netherlands project, ILEIA (Information Centre for Low-External-Input and Sustainable Agriculture). She is the co-author of several books, including ECOFARMING IN AGRICULTURAL DEVELOPMENT, JOINING FARMERS' EXPERIMENTS, FARMING FOR THE FUTURE and PLANNING WITH PASTORALISTS.

*H*ENK DE ZEEUW was educated at the Land Water Development College, Arnhem, after which he gained an MSc in Rural Sociology at the Agricultural University of Wageningen in the Netherlands. After doing some university teaching and research in the Netherlands, he joined the staff of the International Agriculture Centre in Wageningen in 1978, where he coordinated various research and training projects related to rural development. From 1984 to 1987 he worked in Colombia as an agricultural extension expert in the regional development project DIAR and he has visited Latin America frequently ever since. He has been a senior consultant with the ETC Foundation since 1987. His particular interests include Transition to Sustainable Agriculture, Rural Communication, Extension and Training, Participatory Technology Development, and Urban Agriculture. He is the author of numerous publications.

Developing Technology with Farmers

A Trainer's Guide for Participatory Learning

Laurens van Veldhuizen

Ann Waters-Bayer

Henk de Zeeuw

ZED BOOKS LTD London and New York

ETC NETHERLANDS

Developing Technology with Farmers: A Trainer's Guide for Participatory Learning
was first published by Zed Books Ltd, 7 Cynthia Street, London N1 9JF, UK *and*
Room 400, 175 Fifth Avenue, New York, NY 10010, USA in 1997

in association with
ETC Netherlands B.V., Kastanjelaan 5 P.O. Box 64, 3830 AB Leusden, The Netherlands *and*
The Technical Centre for Agricultural and Rural Cooperation* (ACP-EU), CTA, Postbus 380,
6700 AJ Wageningen, The Netherlands

Distributed in the USA exclusively by
St Martin's Press, Inc., 175 Fifth Avenue, New York, NY 10010, USA

Copyright © ETC Netherlands B.V., 1997

Cover and book design: Lee Robinson/Ad Lib Design, London N19
Edited and typeset by Helen Robertson, London N11
Printed and bound in the United Kingdom by Redwood Books, Trowbridge, Wiltshire

The rights of the authors of this work have been asserted by them in accordance with the
Copyright, Designs and Patents Act, 1988

---

*Technical Centre for Agricultural and Rural Cooperation (ACP-EU) The Technical Centre for Agricultural and Rural Cooperation (CTA) was established in 1983 under the Lomé Convention between the African, Caribbean, and Pacific (ACP) States and the European Union Member States.

CTA's tasks are to develop and provide services that improve access to information for agricultural and rural development, and to strenghten the capacity of ACP countries to produce, acquire, exchange and utilise information in these areas. CTA's programmes are organised around three principal themes: strenghtening facilities at ACP information centres, promoting contact and exchange of experience among CTA's partners and providing information on demand.

A catalogue record for this book is available from the British Library

Library of Congress Cataloging-in-Publication Data

Veldhuizen, Laurens van
    Developing technology with farmers: a trainer's guide / L. van Veldhuizen, A. Waters-Bayer, and
H. de Zeeuw.
      p.   cm.
   Includes bibliographical references and index.
   ISBN 1-85649-489-6 (cl). – ISBN 1-85649-490-X (p)
    1. Agricultural extension workers – Training of. 2. Farmers – Training of. 3 Agricultural extension work.
4. Agriculture – Technology transfer. I. Veldhuizen, Laurens van. II. Waters-Bayer, Ann.
III. Zeeuw, Henk de. IV. Title.
S544.V44 1997
630'.71'5–dc21   96-48896
                                                                                                                    CIP

ISBN 1 85649 489 6 hb
ISBN 1 85649 490 X pb

ACKNOWLEDGEMENTS

THIS GUIDE builds on the experience of numerous trainers and practitioners in different parts of the world. It is impossible to mention each by name, but their openness in sharing training approaches and ideas has been the single most important factor enabling us to compile this guide.

A crucial step in the development of the guide was the Workshop on Training Participatory Technology Development and NGOs, held in November 1990 in Leusden, the Netherlands. Representatives from NGOs practising PTD in the field met with experienced trainers and developed an outline for a training guide, which formed the basis of this book. The input of all workshop participants and their organisations is gratefully acknowledged. They included the NGOs CYSD, India; SIBAT, Philippines; ACDEP, Ghana; ENDA, Zimbabwe; IIRR, Philippines; CELATER, Colombia; EDOC, Mexico and World Neighbors, Burkina Faso, as well as trainers and practitioners from the support organisations IIED, UK; University of Hochenheim, Germany; CIAT, Colombia; University of Wageningen, the Netherlands and ETC-ILEIA, the Netherlands.

We obtained occasional feedback on the use of the guide in the years after 1990, but organised a more systematic consultation with users in 1995. Many took the time to answer our long list of questions, gave suggestions for strengthening the guide, and shared new training ideas and materials. Wherever relevant, we have given, to the best of our knowledge, due reference to the sources of ideas presented in the book.

Participatory Technology Development, as we understand it, is not an end in itself but aims at opening up ways towards the sustainable use of natural resources for the benefit of both men and women farmers throughout the world. Our colleague Erik van der Werf screened the draft guide for consistency in this respect and contributed the unit on sustainable agriculture in Part 1. Verona Groverman, another colleague at ETC, helped us very much in taking social and gender differentiation seriously throughout the guide and contributed the unit on gender analysis in Part 1. We greatly appreciated their constructive comments. Stimulating discussions in the St Ulrich group of European-based PTD advocates considerably deepened our understanding of PTD and contributed, either directly or indirectly, to numerous sections of the guide.

We are grateful, finally, to the NOVIB and the Ministry of Development Cooperation, both in the Netherlands, for supporting the publication of this book.

Application of PTD in the field and PTD training efforts continue to generate new insights and new training ideas. In fact, one of the most difficult challenges in preparing this book has been to stop writing and close our eyes to imagine new materials and ideas, which still continue to come in.

We hope that the guide will inspire trainers to maintain this creativity and innovation and to share their experiences both with each other and with us. The six-monthly *PTD Circular*, published by ETC, continues to be open for such exchange.

Laurens van Veldhuizen
Ann Waters-Bayer

Acronyms used in this guide

ACDEP	Association of Church Development Projects
CBEE	Community-Based Experimentation and Extension
CIAT	Centro Internacional de Agricultura Tropical
DELTA	Development Education and Leadership Teams in Action
ETC	Foundation for Ecology, Technology and Culture
FAO	Food and Agricultural Organization of the United Nations
FARMI	Farm and Resource Management Institute
FMD	Forestry Manpower Development Consultants
GO	governmental organisation
GRAAP	Groupe de Recherche et d'Appui pour l'Autopromotion Paysanne
GTZ	Deutsche Gesellschaft für Technische Zusammenarbeit
HEIA	high-external-input agriculture
ICLARM	International Centre for Living Aquatic Resources Management
ICRISAT	International Crops Research Institute for the Semi-Arid Tropics
IDS	Institute of Development Studies
IIED	International Institute for Environment and Development
IIRR	International Institute for Rural Reconstruction
ILEIA	Information Centre for Low-External-Input and Sustainable Agriculture
IPM	Integrated Pest Management
IPRA	Investigación Participativa en Agricultura
IRRI	International Rice Research Institute
ISNAR	International Service for National Agricultural Research
ITDG	Intermediate Technology Development Group
KWDP	Kenya Woodfuel Development Programme
LBL	Landwirtschaftliche Beratungszentrale Lindau
LEISA	low-external-input and sustainable agriculture
M&E	monitoring and evaluation
NAR	national agricultural research
NCCK	National Council of Churches in Kenya
NGO	nongovernmental organisation
NOVIB	Nederlandse Organisatie Voor Internationale Betrekkingen
NWPDZ	North Western Province Dry Zone Participatory Development Project
ODI	Overseas Development Institute
OFCOR	On-Farm Client-Oriented Research
PLA	Participatory Learning and Action
PMHE	Promoting Multifunctional Household Environments
PRA	Participatory Rural Appraisal
PRRM	Philippine Rural Reconstruction Movement
PTD	Participatory Technology Development
RAAKS	Rapid Appraisal of Agricultural Knowledge Systems
SIMAS	Servicio de Información Mesoamericano sobre Agricultura Sostenible
SNV	Stichting Nederlandse Vrijwilligers
SWOT	Strengths, Weaknesses, Opportunities, Threats
TD	technology development
TOT	transfer of technology
UNDP	United Nations Development Programme
UNESCO	United Nations Educational, Scientific and Cultural Organization
UNICEF	United Nations Children's Fund

CONTENTS

INTRODUCTION / AN OVERVIEW	1
PART 1 / BASIC ORIENTATION AND SKILLS	21
PART 2 / TOWARDS AN AGENDA FOR ACTION	85
PART 3 / FARMERS' EXPERIMENTATION	141
PART 4 / SPREADING AND CONSOLIDATING THE PTD PROCESS	189
APPENDIX / REFERENCES, RESOURCES, CONTACTS	215
INDEX	227

AN OVERVIEW

INTRODUCTION

INTRODUCTION / AN OVERVIEW

PARTICIPATORY TECHNOLOGY DEVELOPMENT
- A people-centred and ecological approach — 4

HOW TO USE THIS GUIDE
- Aim and intended users — 5
- Structure and use of the guide — 5
- Sources of materials — 6

CHANGING THE APPROACH
- The change process — 6
- Creating favourable conditions for PTD — 8
- Building external relations — 10

LEARNING FOR PTD
- Basic principles of experiential learning — 11
- Designing the training strategy — 12
- Planning staff training — 15

BOXES
1. NGOs introduce LEISA/PTD in Northern Ghana — 14
2. Starting up PTD in Andhra Pradesh — 15
3. A LEISA/PTD training programme for NGOs in Uganda — 16
4. Tips for effective PTD training — 17
5. Schedule of PTD training for fieldstaff of the PMHE project — 18
6. Schedule of a 4-day introductory PTD training — 19
7. Effective use of plenary, individual and group sessions — 19
8. Basic principles of successful training events — 20

INTRODUCTION

PARTICIPATORY TECHNOLOGY DEVELOPMENT: A PEOPLE-CENTRED AND ECOLOGICAL APPROACH

Throughout the various new fashions that have shaped rural development programmes over the past few decades, increasing emphasis has been laid on the involvement of local people. Some advocates stress this as a means of making programmes more effective and ensuring that activities continue after the programmes have ended. In this book, we go a step further: in our view, the very core of the development process is the increased control that people gain over shaping their own lives. This implies that development workers must get involved – i.e. participate – in the ongoing development efforts of rural families and communities.

Experiences with participatory development approaches have been particularly well documented in agricultural programmes (e.g. Bunch, 1982; Chambers et al., 1989; Farrington & Martin, 1988; Haverkort et al., 1991; Chambers, 1992). Some of these approaches find their origins within research organisations and externally-assisted government development projects, while others have been developed by extension staff of pioneering Non-Governmental Organisations (NGOs). The Institute for Development Studies (IDS) of the University of Sussex collected some of these experiences for a workshop in Brighton, England, in 1987, and the ETC Foundation reviewed these and yet more experiences for a workshop in 1988 in Leusden, Netherlands. The existing term "Participatory Technology Development" (Tan, 1986), or "PTD", was then used to refer to the entire process in which development workers facilitate the generation and dissemination of agricultural innovations together with rural men and women.

PTD is essentially a process of purposeful and creative interaction between rural people and outside facilitators. Through this interaction, the partners try to increase their understanding of the main traits and dynamics of the local farming systems, to define priority problems and opportunities, and to experiment with a selection of "best-bet" options for improvement. The options are based on ideas and experiences derived from both indigenous knowledge (both local and from farmers elsewhere) and formal science. This process of technology development is geared not only towards finding solutions to current problems, but also towards developing sustainable agricultural practices which conserve and enhance the natural resources so that they can still be used by future generations. Most important of all, PTD should strengthen the capacity of farmers and rural communities to analyse ongoing processes and to develop relevant, feasible and useful innovations.

Rural people's capacity to cope with and stimulate change will be of crucial importance if the challenge of raising the level of agricultural production while safeguarding the land is to be met. It has become increasingly evident that scientists alone cannot generate site-specific technologies for the wide diversity of conditions of resource-poor farmers throughout the world, or even within one country (Reijntjes et al., 1992). This is not only true of the more marginal areas where agriculture based on high levels of external inputs could not take hold, but also in those better-endowed areas where efforts are now being made to reduce the use of external inputs, for economic or ecological reasons. In both cases, the knowledge and skills of farmers in, for example, influencing soil fertility or managing pests and diseases, will play a key role in developing appropriate technologies.

The process of technology development is closely linked with a process of social change. Encouraging local innovation through self-organised planning, implementation and evaluation of systematic experiments fosters self-respect and self-confidence in the rural communities involved. It also fosters a process of cultural awareness and change, as the planning and assessment obliges the participants to take account of their situation and the responsibilities of different people in the community, e.g. the different needs of men and women and the different barriers they face in trying to change their situation. Moreover, each community is made up of a variety of individuals with a diversity of interests and lifestyles, as becomes apparent especially when criteria for "improved" technologies are debated. Yet another complication may be that the natural resources are being used not just by one sedentary community but also by temporary migrants and new immigrants into the area. Therefore, the process of local technology development in agriculture will also often involve negotiations and the development of new institutions to accommodate change in the use of natural resources.

HOW TO USE THIS GUIDE

AIMS AND INTENDED USERS

The aim of this book is to provide support to rural development organisations that wish to become involved in PTD or to strengthen their present activities in participatory research and development. These may be government extension services, public or private research institutes, Non-Governmental Development Organisations (NGDOs) or community-based organisations. The book is addressed primarily to the people within these organisations who are responsible for staff development and who plan and carry out training activities. They are referred to here as "trainers".

The book provides trainers with a set of resources that can be used in preparing fieldstaff for PTD. Its intention is to facilitate the design of local PTD training events and to stimulate the development of training aids adapted to the local context, needs and resources. It is not intended to be a manual for a "standard" course in PTD, and does not suggest a predetermined course outline. Using this training guide in such a way would go against the very essence of a participatory development approach and the learning results would probably be disappointing.

The didactic focus of the guide is on experiential and problem-solving learning. Participants in the training events are meant to play an active role, linking the development of new insights with their existing experiences and stimulating each other to reflect critically on new ideas in the context of those experiences. The book should help trainers to develop ways of changing the attitudes of their fieldstaff and nurturing the skills that are needed to interact with farmers in technology development.

The training activities are primarily oriented to fieldstaff with a certificate or diploma level of agricultural education. It is assumed that they are already working in the field with farmers, but have limited experience in investigatory research or experimentation. Feedback received on earlier drafts of this training guide suggests that, with a few exceptions, the materials presented will need to be considerably modified if they are to be used in training farmers and farmer-leaders or in training development programme managers and policymakers.

The guide focuses on PTD in agricultural production. However, the realisation is growing that PTD concepts and methods are also relevant in other fields such as food processing and small-scale industry (van der Bliek & van Veldhuizen, 1993). Development organisations should feel free to use the guide to develop PTD training suitable for these and many more fields.

STRUCTURE AND USE OF THE GUIDE

After this introductory section, the guide consists of four parts which present training approaches and materials closely related to important phases in the PTD process.

This **Introduction** includes an analysis of the role of training in the wider context of promoting a PTD approach within an organisation, as well as a discussion of important institutional conditions for successful PTD programmes. It also presents important issues to consider in planning, implementing, monitoring and evaluating PTD training programmes.

Part 1 focuses on the basic orientation and skills that fieldstaff need if they are to be successfully involved in PTD activities. This includes a critical reflection on previous experiences in working with men and women farmers, a comparison of PTD with other approaches to technology development, and an assessment of the role that PTD may play in the search for sustainable forms of agriculture. The main principles of the PTD approach and its overall framework are presented. This part ends with a discussion of the fundamental skills in listening and probing which are needed for dialogue with men and women farmers, and suggests how these skills could be enhanced.

Part 2 covers joint analysis and planning, presenting issues and methods related to the first phases of the PTD process. This includes establishing contacts with farmers and communities, analysing local problems and opportunities together with them, and stimulating the farmers – if need be, in separate groups of men and women – to select promising options to try out.

Part 3 builds on the recognition that many farmers perform their own small experiments as part of a process of gradually changing their farming system. Different ways in which fieldstaff can support farmers' experimentation are presented. Special consideration is given to the role of farmers' groups in local experimentation and innovation, and to the joint monitoring and evaluation of the experiments by farmers and fieldstaff.

Part 4 focuses on the last phase of the PTD process, in which the outcome of experimental activities by and with farmers is spread to other farmers. In this extension phase, farmers themselves again play a crucial role, not only in suggesting new ideas they have tried themselves but also in showing other farmers how they, too, can experiment with such ideas. This final section also discusses ways of ensuring that farmers' groups and communities can continue to practise PTD in addressing other present and future issues and can draw upon agricultural services to support their efforts.

These four parts are divided into *learning units*, each focused on a particular subtheme and with specific learning objectives. An overview is given of the main concepts and contents of each unit, and possible learning activities are suggested and described. In most cases, the unit contains elements intended to facilitate learning in terms of concepts, attitudes and skills.

The **Appendix** at the end of the book gives a complete alphabetical list of references. In addition, it contains information useful when planning or implementing PTD training programmes – details of audiovisuals and where to obtain them, sources on participatory training, PTD networks and contacts, and periodicals regularly featuring PTD contributions.

It must be stressed once again that this guide is not meant to be a manual for direct use in training PTD. It should rather be seen as a source of *ideas*. Trainers are challenged to reflect critically on the concepts and the learning activities described here, and to select and adapt the most relevant ones for their situation. In designing their own training course with the aid of this guide, trainers will have to take into account their own experience and capabilities, the current operational methods of their organisation, the skills and attitudes of staff who will be involved in the training, and the experience that staff may already have in participatory research and development.

Learning activities will be most meaningful if they build on local experiences. The learning process then becomes related to real farmers' groups and farming situations in the area where the participants are working. A final challenge to users of this guide is to ensure that training experiences are carefully documented and monitored. This will not only help the trainers to improve the quality of their work but will also assist in developing locally-adapted training contents, methods and materials which can be used by other trainers.

SOURCES OF MATERIALS

This guide builds on the experiences of numerous trainers working both within and outside PTD programmes. A first draft was compiled on the basis of the results of the international trainers' workshop held in Leusden in 1990. That draft, which included well-proven as well as newly developed training ideas, was then used and adapted by many trainers in Asia, Africa and South America. Feedback was later collected from these trainers, particularly about their adaptations of training ideas and development of new ones suitable for inclusion in the book. To the best of our knowledge, we have given due reference to all trainers, authors and developers of the training ideas presented in this guide. Their addresses are given in the Appendix at the end of the book, so that they can be contacted directly.

Thus, although the initial draft version of the training units included here have been tested, some of the proposed learning activities which have been incorporated into this revised version are simply new and stimulating ideas that have emerged from reflection on experience. However, all PTD methods and concepts included in the units have been developed in the field.

CHANGING THE APPROACH

If a training programme in PTD is to be successful, it must be part of a wider process of adjusting an organisation's approach to accommodate PTD. Here, we outline such a change process and indicate some preconditions for making the changes effective.

THE CHANGE PROCESS

To begin with, there must be an awareness at some level within the organisation that it needs to adopt a more participatory approach. Rather than simply jumping into staff training, an organisation should try to make a thorough preparatory analysis in order to seek answers to at least the following questions:

- What are the major shortcomings in our present approach and operational methods, taking into account the agroecology in our working area and the major needs and constraints of the men and women with whom we work in the rural communities?

- How would PTD fit into our present approach? Would it alleviate some of the current shortcomings?

- What kind of adaptations have to be made in the way we work so that PTD can be included?

- Who within our organisation should be involved in further analysis, planning, staff training, monitoring and evaluation, and the documentation of PTD activities?

- What changes do we need to make in our internal organisation in order to be better prepared to handle PTD programmes?

- What is the scope and need for cooperation concerning PTD with other organisations in our region?

- What kind of specific assistance do we need from other agencies?

- Which donor organisations may be willing to support PTD activities?

The preparatory analysis may include such activities as a management meeting, discussions with fieldstaff, a workshop with key resource persons, and a staff meeting to discuss the analysis. This may begin with only one, or a few, of the staff making a preliminary analysis, which then leads into a more thorough analysis involving other levels within the organisation and possibly other organisations working in the same region. At this point, a joint workshop may be suitable.

Further information on networking around issues of PTD in agriculture can be found in Alders *et al.* (1993). If the organisation indeed agrees that a PTD approach is desirable, the analysis will then lead to initial actions, mostly probably in one of the following four areas:

1. Some elements in the organisation's field methodology may be identified that can be improved immediately. For example: a checklist used in discussions with farmers on agricultural problems may be adjusted to include items referring to farmers' knowledge and experiments; farmers may be invited to take part in assessing trials carried out by the organisation; or more attention may be given to discussing research needs with representatives of informal and formal farmer organisations.

2. Apart from these incremental changes, a decision may be made to set up relatively small-scale pilot activities with a PTD approach, involving only some highly-motivated fieldstaff. In this way, some basic experience can be gained before involving the entire organisation in widescale change. A favourable response by farmers in these initial activities will increase the motivation of fieldworkers and the organisation, whereas problems stemming from overambitious and premature widescale implementation could lead to frustration. The pilot activities are useful as a learning experience for testing and adapting the methods applied by the fieldworkers, and in developing training materials. They can also provide a kind of "resource centre" in the field as a basis for expanding activities into other areas. Large governmental organisations with the mandate to cover an entire country may find it difficult to alter their approach in only one region or district, even if only for the limited period of a pilot phase. If they feel obliged to operate uniformly nationwide, yet do not want to risk a complete change in approach and an upheaval in structure, they will have to accept a slower learning process. This will mean implementing very small changes throughout the entire system, as gradual steps towards further-reaching changes, over a period of several years.

3. A third direction for initial action which may emerge from the preparatory analysis is staff training. This is the direction on which this book concentrates. Later in this introduction, the planning of training and appropriate training strategies is discussed in more detail.

4. The fourth direction for action – organisational change – is discussed in the following section, in which important preconditions for PTD are outlined.

Through its initial actions, the organisation will gain worthwhile experience and insight into indigenous knowledge, farmers' experimental practices, the adequacy of their experimental methods, the appropriateness of certain technologies, and the need for new information and/or linkages. However, if the full benefit of learning is to be gained from these initial actions, they need to be monitored, documented, discussed and analysed. This puts even greater demands on staff already overburdened with trying to cope with day-to-day operational problems. A compromise must be made between completeness of data, depth of analysis and feasibility. Unit 4.B (pp. 207–13) of this training guide, which deals with sustaining the PTD process, gives some guidelines and training ideas on how such a compromise could be reached.

Involving programme coordinators, subject specialists and fieldworkers in analysing field experiences helps to motivate the fieldworkers and greatly enhances the insights gained at all levels. A series of SWOT sessions (analysis of Strengths, Weaknesses, Opportunities and Threats) is an example of a simple but powerful tool that can be used to analyse jointly the experiences, develop appropriate solutions, and adapt strategies and methods to suit the local situation. Applying such participatory methods within one's own organisation contributes directly to developing appropriate attitudes and skills for work with rural communities. At this point, the organisation will be starting gradually to intensify and scale up the PTD process with farmers. Careful preparation for wider application will then be needed.

CREATING FAVOURABLE CONDITIONS FOR PTD

To accommodate a PTD approach, adjustments will usually be needed both within the organisation and in its relations with other organisations. These adjustments will help to create conditions that allow the fieldstaff to apply what they have learnt in PTD training. The list of issues presented here is far from exhaustive; it is intended rather to stimulate thinking about conditions that favour PTD.

The importance of intra-organisational issues is well expressed by Ueli Scheuermeier: "PTD starts within an organisation. Only once that works, and the fieldworkers are listened to and learnt from, can we expect to get ahead with listening to villagers and jointly get cracking" (pers. comm., 1990). Besides this central point of taking a participatory approach within the development organisation itself, some other favourable conditions are:

- **Flexibility in programming:** The main focus of the PTD approach is to start with expressed farmers' problems and try to solve them. Therefore, the organisation should have the flexibility to respond to a variety of problems and issues raised by different categories of farmers. There is, however, a danger that too much openness and flexibility may lead to poor coordination, loss of special knowledge and skills, and poor allocation of resources (see "Allocation of resources", p.10). It is at least desirable that the organisation should be able to help farmers make links with other sources of support to follow up on the problems identified.

- **Decentralisation of decision-making in planning:** Day-to-day decision-making within field programmes needs to be decentralised so that fieldworkers can be flexible and efficient in their interactions with farmers. As they have an intimate knowledge of the actual field situation, fieldworkers should participate in higher-level planning of the field programme.

- **Regular evaluation of activities and impact:** Working in partnership with farmer communities cannot follow a clear-cut blueprint. Any organisation involved in PTD needs to realise that it does not have *the* answer to farmers' problems; it must be prepared to learn through its interaction with the farmers. To be able to learn, the organisation must periodically examine the way it is working with farmers, possibly by means of annual evaluation workshops. Adequate evaluation can be made only if relevant information is collected to document what has been done and how effective it was. This should include assessment of the extent to which different types of farmers (e.g. men and women, different wealth classes, different ethnic groups) are participating in or are affected by the PTD activities.

- **Re-assessment of the roles of fieldstaff and co-ordinators:** The conventional roles of fieldstaff and coordinators of development programme will require re-assessment. The coordinators will have to work with the fieldstaff in the same way as they expect the staff to work with farmers – facilitating, supporting, filling in gaps, etc. Meetings between fieldstaff and higher levels within the organisation may have to be organised to give an opportunity to discuss problems encountered in the field and to

To be effective, PTD needs a participatory approach within the development organisation

develop solutions together. Even more important is cultivating a general attitude among management staff of being approachable by fieldworkers and listening to the problems expressed by both fieldworkers working with men and those working with women farmers.

■ **Systematic staff development:** The effectiveness of fieldstaff is crucial in implementing PTD approaches. Staff development therefore needs special attention. Some questions that need to be considered are:

- What criteria should be used in selecting future fieldstaff: agricultural training, community organisational skills, gender, experience in working with both men and women, or some combination of these criteria plus others?
- Where should potentially good staff be actively sought?
- What kind of initial training and guidance should be given to new staff?
- What kind of in-service training is needed, and when?

- What are the possibilities of sending staff to other NGOs or elsewhere for relevant training?

■ **Discovering new technical options:** Farmers involved in a PTD process will look to the support organisation for suggestions on new technologies worth testing. The organisation must therefore have a strategy of keeping itself informed about new ideas of possible interest to farmers and actively seeking possible explanations for what farmers are finding in their experiments. These innovations and explanations may come from the same region or from other regions with similar conditions, and from either formal or informal research.

■ **Storage and use of information:** Once an organisation has become involved in a PTD process with farmers, a vast amount of information will be generated about, for example, the local agroecological situation and changes, the indigenous knowledge and experiences of farmers, the interaction between fieldstaff and farmers, and, of course, the results of the experiments and investigations performed together with farmers. Choices must be made about what information to store and how to store it, how to analyse the data and how to use the results to improve the organisation's performance. Without a good filing system, everything will be lost when a few experienced staff members leave the organisation, or cupboards will be filled with unused data (Jiggins & de Zeeuw, 1992). Appropriate information and documentation services, such as "minilibraries", should be available not only to the development organisation but also to the farmers.

- **Allocation of resources:** A development organisation needs to allocate its limited funds and staff carefully in order to use them most efficiently. Becoming more involved in PTD approaches will normally require a change in resource allocation: e.g. more emphasis on staff training and field operations and less on office activities and structures. A PTD programme may also require an unspecified fund to cover unforeseen risks and to support small projects that emerge from the experimentation.

BUILDING EXTERNAL RELATIONS

Most organisations will not be able to implement PTD approaches without cooperating with other agencies. Especially an organisation commencing PTD will need support from others with more experience. But also, in the ongoing PTD process, various organisations will have different but closely interactive roles to play:

Agricultural research institutes:
- providing information on new technologies;
- participating in fieldwork during situation analysis and the identification of "best-bet" options;
- advising on design and monitoring of farmers' trials;
- carrying out on-station research into field-generated innovations or adaptations or into questions raised by farmers;
- providing specialist services (e.g. entomology or virology); and
- using their knowledge of farmers' situations and questions to influence the national research agenda.

Government extension services:
- providing technical advice about specific technologies;
- preparing training/extension materials on those technologies;
- providing secondary data (soils, climate, prices, etc);
- encouraging farmer-to-farmer extension and sharing of results in a wider area; and
- using their knowledge of farmers' situations and questions to influence government extension policy.

NGOs or development projects:
- mobilising farmers and giving them organisational support;
- facilitating farmer-led situation analysis and planning of the local research agenda;
- supporting farmers in carrying out their experiments;
- giving guidance in dealing with gender issues;
- mediating in conflict resolution; and
- encouraging farmer-to-farmer extension.

Farmer organisations:
- identifying and articulating felt needs and current problems;
- coordinating the process of carrying out, monitoring and evaluating farmers' experiments;
- organising farmer-to-farmer visits and exchange, both within their own area, and further afield; and
- eventually assuming responsibility for continuing the PTD process in the area.

In practice, the distinctions are not always so clear. Overlaps occur, and one type of organisation may expand to fill gaps. In some cases, still other organisations may play an important role in agricultural technology development, e.g. sales agents of farming inputs, manufacturers of tools and equipment, or credit institutions. Periodic "stakeholder" analyses are necessary to ensure that all relevant agencies are involved in the PTD process (see Unit 2.B, pp. 102–12).

LEARNING FOR PTD

The concept of "staff training" varies greatly from organisation to organisation. Indeed, some development practitioners regard "training" as having a "top-down" connotation. We understood it as a learning process in which the participants – assisted by the trainers – are involved in activities which help them to discover how they can improve their performance or to prepare themselves for a new role or task.

The emphasis is on active learning by the participants, rather than passively receiving information from other people. The learning process may include a number of focused training sessions, periods of work in the field and various other learning moments, in an upward spiral of alternating action and reflection. When fieldworkers are being prepared to enter into a PTD process with farmers, the training events are based primarily on the field experiences of the participants and relate to what they encounter in their day-to-day work.

BASIC PRINCIPLES OF EXPERIENTIAL LEARNING

PTD training is based on the same basic principles as the PTD approach itself: experiential learning in an iterative process of action–reflection–action. If, after the participants have gained positive experiences during training, the trainer points out the parallel with the interactions between fieldworkers and farmers in PTD, this can be a powerful learning moment. It is even more powerful if the participants themselves recognise the common aspects of adult learning in their own training and in their work with farmers in the field.

Experiential learning is based on exchange, analysis and systematisation of the participants' own experiences. Two types of experience are involved:

- *Learning from real-life experiences* means starting with the participants rather than the "teacher" – with the experiences they have gained in their home and work situation, with their analysis of causes and possible solutions to the problems at hand. It does not start with "theory" and knowledge gained from books. This parallels the recognition in PTD that the knowledge of extensionists or researchers complements that of farmers.

- *Learning from systematic reflection* on what the participants are doing and experiencing here and now in the training group. How did we cooperate, solve problems, take decisions, handle conflicts – here in this group – and what can we learn from this? The trainer creates opportunities for situations which allow experimentation with certain types of behaviour. The participants learn by reflecting on these experiences.

Experiential learning implies that the trainer arranges learning experiences through which participants can discover and develop new insights and skills themselves, rather than being taught them by the trainer. As a result they will leave the training not with a PTD blueprint, but rather with better insight into how they can support technology development by farmers in their own situations.

The learning process follows an upward spiral of action, reflection and action and includes the following phases (based on Kolb & Fry, 1975; adapted by Lammerinck & de Zeeuw, pers. comm.):

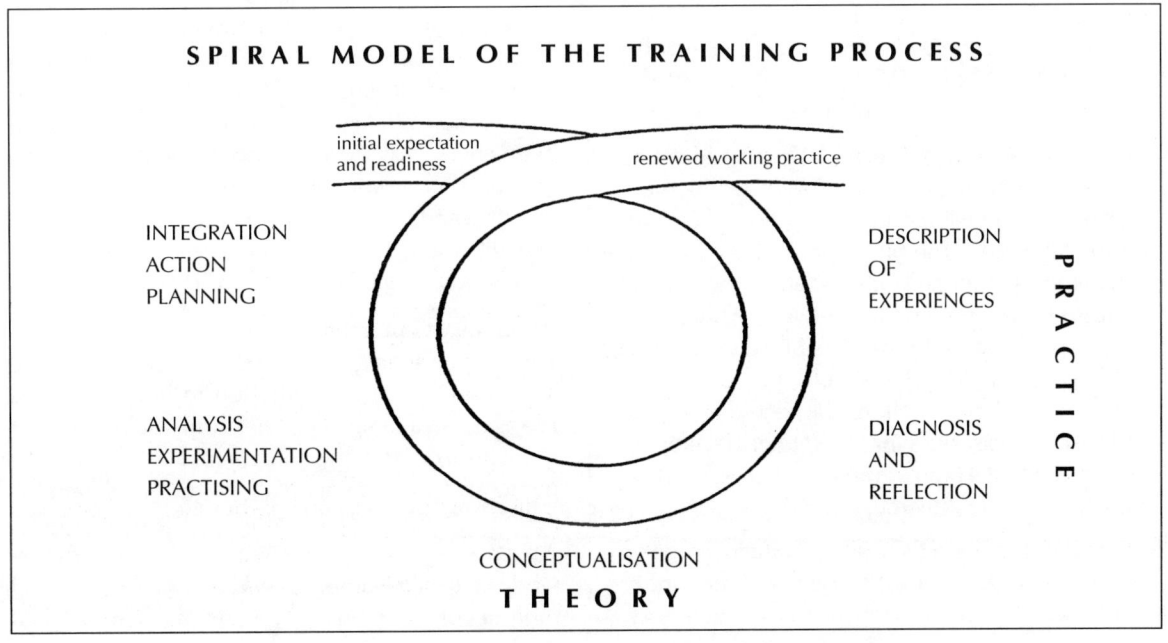

1. **Orientation:** The learning process begins with a clarification of *what* the training is all about. This can be facilitated by showing participants a "typical" case of the subject – either orally, on paper, in a film, with slides or in a socio-drama. This should lead to a discussion on *why* it is important to discuss the subject. What do we think the situation should be? Why do we want to develop our knowledge and skills in this field? This will also help participants to develop a clear and shared understanding of *how* the learning process will be organised.

2. **Generation of participants' experiences:** Participants describe and exchange their individual experiences in this field and/or with this problem: What did each individual observe/experience? Where, when, how? Under what conditions (physical, institutional, social) did it happen? What consequences did they observe? How do they handle such a situation normally? What did they or others do to solve the problem, and how did it work out?

3. **Reflection on the experiences:** Participants compare the emerging picture of the actual practices with their views on how the situation should be. What is the gap between what we do and what we think we should do? The trainer helps the participants to structure and compare their experiences, and to analyse the main differences and convergences in their practices. What are the major aspects of the problem/subject? What causes can be identified, how are these interrelated, and under what conditions do they take effect? Why did this solution work here and not there?

4. **Conceptualising and formulating the learning tasks:** The emerging pieces of knowledge are grouped into logical and interconnected clusters and "labelled". The participants then draw up hypotheses (informed guesses) about the general causes of the discrepancy between what is and what should be, and about possible solutions. This helps them define what should be dealt with in greater detail in the training – aspects that require further analysis, skills that need to be developed, or aspects of the problem about which more information must be gathered.

5. **Focused learning activities:** The participants delve deeper into the subject, carrying out the further analyses they have identified as being necessary, practising desired skills, gathering additional information about particular aspects of the subject and actively seeking potential solutions. They carry out a critical review of selected texts, analyse case studies, do fieldwork, interpret available records, conduct and evaluate small experiments, or perform similar activities which the trainer feels will help develop the required skills.

6. **Integration and translation to the work situation:** The main findings of the learning process are brought together and reviewed for their relevance to and feasibility in the specific situation of individual participants.

This learning process provides the basic structure for the whole training event, but also for each learning unit, and even within specific learning activities.

DESIGNING THE TRAINING STRATEGY

Before planning a specific training event, a general training strategy needs to be developed: How are we going about increasing the staff's capacity in PTD? Some of the major choices made, either explicitly or implicitly, when designing a training strategy include:

Individuals vs teams vs "task force"

The training strategy may be focused on individuals with a certain type of task from different parts of the organisation, or on a team working together on a certain project or area. A third option is to include in the training process all those who have something to do with the task at hand, both within and outside the organisation. Strategies that focus on training a team or an entire "task force" are normally more challenging and rewarding, but are also more complex to organise and manage. As PTD processes are highly interactive and involve various types of actors, a team or "task force" approach is often advisable.

Temporary vs permanent

The training strategy may focus strongly on creating a temporary learning situation to induce certain changes, i.e. a seminar or a field trip. Alternatively, it may aim at organising a continuous learning process: participants meet periodically to review certain experiences and study further topics of interest to them. A participatory monitoring system can create a framework for regular learning. A system of peer review and mutual consultation can be organised. Individual "backstopping" support to participants can facilitate ongoing learning in the field. In the case of PTD, the development of permanent learning systems needs major attention, even though this is more expensive and more difficult to set up than short-term training. Efforts should be made to identify and build on learning mechanisms already within the organisation. Box 1 presents the experiences of an NGO network in Ghana with such a longer-term learning process.

Trainer-controlled vs learner-controlled

Training activities can either be planned and shaped mainly by the trainers, more or less according to directives coming from higher levels within the organisation, or the participants themselves can be strongly involved in shaping the training. Learning about PTD will be much more effective when participants are involved in planning and implementing their own learning process. The systematic sharing of and reflection on their experiences in such involvement will contribute to learning about PTD itself.

Field-based vs centre-based learning

The emphasis may be on learning from concrete practical experiences, e.g. through joint fieldwork by more experienced and less experienced workers, or through joining a farm family in their daily life for a period of time. Alternatively, the participants may be brought together in a centre to study certain topics intensively for several days. The field-based approach is normally more effective in developing appropriate attitudes and skills for PTD. The centre-based approach offers better conditions for systematic reflection, as it gives some distance from day-to-day practice. The participants can then concentrate on new concepts and experiment freely with new methods. But even in centre-based courses or workshops, one can (and should!) practise problem-oriented "learning by doing", e.g. by combining theory with practical exercises and making extensive use of case studies, role plays, simulation and communication games. A combination of centre-based activities with fieldwork will be most effective. The training programme with the Department of Animal Husbandry in India (Box 2) gives a good example of such a combined approach.

Focus on the "P" vs the "T" in PTD

Training in which participation is the crucial issue will concentrate on clarifying the concept of participation and developing relevant attitudes and skills, such as respect for farmers' knowledge, skills in dialogue and stimulating creative interaction, awareness of and skills in dealing with gender issues, and knowledge of participatory research and extension methods. Strategies emphasising the "P" will often include "laboratory"-type training methods as well as real-life interaction with farmers.

Alternatively, if the focus is on developing technologies for sustainable farming systems, PTD training will be designed to enable staff to support such activities with farmers. It will focus on changing the way fieldworkers regard nature and agriculture, enhancing their understanding of the principles and techniques of ecological farming, and developing skills to observe (non-) balances in agroecological systems and to experiment with alternative agricultural practices. Strategies focusing on the "T" will use methods like field observations of ecological processes, field tours to sites where experimentation with ecological farming methods is taking place, and "learning by doing it yourself".

The most effective training strategies will often be those which combine the two dimensions: introducing PTD not only as a new approach to research and extension, but also as a new way of looking at nature and agricultural development. An example of a one-year training programme for NGO staff in Uganda that creatively incorporated both dimensions is given in Box 3, p.16.

Direct training vs training-of-trainers

Within an organisation, a learning process can be developed directly with the fieldworkers. Alternatively, initial training may involve only the field-team coordinators, who will in turn train their teams, with or without support from the initial trainer. Another possibility is to develop a cadre of trainers who, in turn, will train field-team coordinators and/or fieldworkers, possibly in a number of different organisations. In the short run, the first option will be the most effective, but it is also the most limited in terms of outreach. The last two options allow continuation of the learning process over a longer period of time and enable a wider spread of the training activities, although the quality of training is likely to be less uniform.

Farmers as "targets" vs trainers

In the training of fieldworkers, farmers may be given a purely passive role, perhaps only being mentioned in discussions as the "target group", or they may be more directly involved in field exercises as "informants".

BOX 1 NGOS INTRODUCE LEISA/PTD IN NORTHERN GHANA

The process in Northern Ghana began with individuals in church-related agricultural projects belonging to the ACDEP network, who were concerned about the current approach and technologies advocated. They started to gather information and, in the process, one of them learned about PTD. As a result, the following sequence of activities took place:

INITIATION (early 1989)
The Chairman of ACDEP raised the issue for discussion at the ACDEP annual meeting. A decision was made to organise a joint workshop in order to incorporate PTD into the projects of network members. Funding arrangements were worked out.

PREPARATION (mid–end 1989)
"Station managers" (project coordinators) individually analysed the agricultural situation and processes in their working area and their own methods of work, with the help of a checklist.

The ACDEP training committee prepared a workshop (1 week): External resource persons assisted the committee in such a way that those involved felt confident to moderate the planned workshop themselves: by using participatory methods during the planning of the workshop, and by trying out/rehearsing what would be done in the workshop.

INTRODUCING PTD (end 1989 – early 1990)

- 1-week **Managers' Workshop** held, to:
  - revive participants' understanding of agricultural production and technology development in the region;
  - develop operational methods for starting PTD processes;
  - enhance commitment and plan for field applications.

  The main subjects and training methods chosen included:
  - an introduction to the basic principles of LEISA;
  - identification of promising technologies (brainstorming);
  - determining steps in developing technologies with farmers (group analysis of their own work methods and possible improvements, and comparison of on-station research, traditional extension, PTD);
  - development of operational methods per "phase" in the PTD process (groupwork);
  - design of a training programme for the fieldworkers in the various projects (groupwork);
  - participatory training techniques (groupwork/tryouts);
  - follow-up planning:
    - PTD as a fixed item on the agenda of the ACDEP annual meetings and agreements to form PTD network;
    - development of format and guidelines for reporting on field activities in PTD and training fieldstaff.

- **Fieldworkers gathered information on farmers' traditional methods of production** and related knowledge and skills, decision-making, organisational systems, experimentation and innovation, with the help of a checklist.

- **Fieldworkers' training workshops per project** (1 week), following the set-up of the managers' workshop. With respect to technologies, each station focused on those most relevant in its own area.

STARTING UP PTD

- **Farmer orientation sessions** (March–April 1990): introduction of the idea of a farmer experimentation programme in selected villages, preliminary analysis, cooperation agreement/arrangements. In cooperation with local leaders.

- **Farmer-designed workshops** (April–May 1990): small groups of farmers analysed priority problems more deeply and made an inventory of options to try.

- **Follow-up training for fieldstaff:** additional training workshops with managers and fieldworkers on the organisation and techniques of farmer experimentation and monitoring, and issues related to stimulating farmer-to-farmer exchange (e.g. cross-visits). The experiences in the foregoing period were discussed and the set-up of the experiments was reviewed.

- **Start of the experimentation with groups of 10 farmers**, comparing the HEIA techniques normally recommended with both traditional practices and new LEISA techniques (from May 1990 onwards).

Sources: Millar et al. (1989) and Millar (1990)

BOX 2 STARTING UP PTD IN ANDHRA-PRADESH

In 1993 the Department of Animal Husbandry went through a review process of their extension approach. The role of "participatory extension" was explored and PTD was identified as one of the crucial processes in this. A series of activities was then decided on to introduce PTD in the Department – not through mere training but by actually starting PTD in the field. More specifically, the activities had the objectives:

- to initiate PTD in a very practical way in one of the project areas, as a first experience on which to base further attempts in other areas.

- to start to specify in more detail the back-up system required for developing PTD as a viable field programme.

A series of three workshops took place as follows …

Date	Workshop	Activities
16 February	**INTRO TRAINING** PTD basics (at Centre)	■ What is PTD? How does it work? Why PTD?
17 February		■ Train the required skills, get a feel for it
18 February		■ Design PTD for animal husbandry in Andhra Pradesh
19 February		
20 February		
21 February	**START-UP FIELD WORKSHOP** Valigonda, doing the real thing	■ Get acquainted with area, public meeting
22 February		■ PTD interactions with villagers
23 February		■ PTD interactions with villagers
24 February		■ PTD interactions with villagers
25 February		■ PTD interactions with villagers, public final meeting
26 February		■ Setting up PTD procedures and plans for Valigonda cluster
27 February		
28 February		
1 March	**REVIEW WORKSHOP** Follow-up and organising	■ Review experiences
2 March		■ Organise operational backing for PTD in the clusters: task-force and resource base
3 March		

Source: Scheuermeier & Sen (1994)

A third possibility is that farmers are involved as actual "trainers". Such a reversal of roles has a positive learning effect.

It must be remembered that creating an open, stimulating and supportive environment within the organisation will be as important for successful PTD as staff training as such, whatever the strategy chosen (see Box 4, p.17).

PLANNING STAFF TRAINING

Planning staff training starts with formulating the overall training programme. This involves three main steps, answering questions such as those following:

■ **Defining the BASIS for the training programme**
What are the main aims of the training? Why is it needed? For what target group(s) is the training intended? What resources are available? Who will contribute to the programme? How will responsibilities be divided between all parties involved? What provisions need to be made for organisational support and follow-up?

■ **Analysing the LEARNING NEEDS**
What are the tasks and situations for which the participants are being prepared? What abilities do they need for this? What knowledge, skills and attitudes? Which of these do they already have, and which are lacking? How do the intended participants define their learning needs? What are the main priorities for the PTD programme? Unit 1.F (pp. 67–74) discusses in detail the various roles of fieldworkers in PTD. Thinking about these roles will help in identifying the required capabilities. The Appendix

to this guide (pp. 217–25) includes several training-of-trainer manuals that discuss training needs assessment tools. In practice, assessment does not stop when the training begins but demands continuous attention throughout.

- **Determining the TRAINING STRATEGY**
What are the assumptions about education on which the training programme is based? Is the focus more on the "P" or on the "T"? What combination of training events (type, location, phasing, etc) and supporting activities will help to realise the desired changes most effectively? (Refer to the issues about training strategy and the various examples discussed above.)

Each training event included in the programme then needs to be planned. This will involve many of the following elements:

- **Developing a PLAN OF TRAINING ACTIVITIES**
Define the learning objectives for the training event. Define the main issues and content. Select relevant existing training modules and units that can be included. Review training experiences and consult key

---

BOX 3   A LEISA–PTD TRAINING PROGRAMME FOR NGOs IN UGANDA

RATIONALE
Ugandan NGOs working as partners of NOVIB and SNV — both Dutch development organisations — held a two-day brainstorming workshop in 1993 on sustainable land-use. It identified a need for training in sustainable agriculture and participatory extension methods for fieldstaff.

Consultation with resource people led to the formulation of a longer-term training programme involving both fieldworkers and managers. This was set up as an iterative process of learning from theory and practice, implementation and work assignments with farmers in participants' own working situations, and the subsequent guided evaluation of experiences.

Timing	Activity	Topics	Comments
March 1994	Pre-workshop assignment	Trends in agriculture locally	Analysis to be completed prior to workshop
April 1994 2 weeks	Introductory workshop LEISA–PTD for 16 extension staff	LEISA principles, soil and water management, PTD basics	
May–June 1994	Field assignment with farmers Two follow-up visits to each organisation	Recognising LEISA, soil fertility and conservation, participatory problem analysis	Each to select 3 out of 5 possible assignments (mostly the technical ones chosen)
July 1994 2 weeks	Second workshop with 22 extension workers and 1 manager	Evaluation assignments, crop management, animal husbandry, problem diagnosis, farmer experimentation, monitoring and evaluation	All participants of first workshop but one returned for this workshop Review of assignments very much appreciated
Aug–Nov 1994	Assignments and follow-up visits	Various LEISA technologies, participatory problem analysis, trends in agriculture	4 out of 7 assignments to be selected Work on assignments better than first round
Dec 1994 1 week	Final workshop with 17 extension workers and 10 managers	Evaluation of assignments, action planning, further networking	Joint work by policymakers and fieldstaff of each organisation

EVALUATIVE COMMENTS:
- The advantages of this iterative training approach easily outweigh the additional labour and costs in comparison with one-off training events.
- In all activities, the same staff from each organisation should participate to ensure the building-up of learning experiences. This needs to be stressed from the outset.

Sources: v.d. Werf (1994) and Bokkestijn & v.d. Werf (1996)

informants about appropriate additional methods. Adapt learning units and prepare new ones, if required. Outline an appropriate structure, sequence and timing for the training event (compare the examples given in Boxes 5 & 6, pp.18 &19) . Prepare each of the planned learning activities (games, cases, role plays, group assignments, inputs, printed materials, audiovisual media) and pretest techniques and materials, where necessary (compare the guidelines in Boxes 7 & 8, pp.19 & 20). Select methods for participatory evaluation of the learning process during the course of the training. Plan the end-of-course evaluation.

- **Selecting and preparing PARTICIPANTS and TRAINERS**
Define selection criteria for participants. Is there a need to give special attention to ensuring participation by female staff? Consider the motivation of participants. Communicate with participants on their training needs and practical issues. Select and prepare trainers and resource persons.

- **Preparing the LOGISTICS**
Arrange for board and lodging, transport, invitations, documentation, classroom facilities, audiovisual and other media, excursions and fieldwork.

BOX 4 TIPS FOR EFFECTIVE PTD TRAINING

- Allow contrast, insecurity and a degree of confusion, as these can be powerful learning tools.
- Encourage role reversals to strengthen the experience of learning about participation: farmers who train extension workers who train programme managers and subject specialists, NGO staff who train representatives of donor organisations and government institutions.
- Use existing PTD programmes as learning environments for newcomers.
- Involve local people in the training as much as possible, and constantly consider how the farmers' role as facilitators can be enhanced.
- Give specific attention to the skills and knowledge of women vs men farmers.
- Allow for using indigenous ways of communication and informal contacts.
- Stimulate fieldworkers to broaden their information base and their perspective about what is happening in other regions, and about topics such as land tenure, markets and government policies.
- Encourage creativity.

BOX 5 SCHEDULE OF PTD TRAINING FOR FIELDSTAFF OF THE PMHE PROJECT

TIME	Monday	Tuesday	Wednesday	Thursday	Friday
7.00 am		BREAKFAST			
8.00 am		Overview of yesterday's session; presentation of the problem tree and discussion	Overview of yesterday's session; presentation of the fieldwork results and discussion	Overview of yesterday's session; **7.** Main theme: "Implications for Implementation"	Overview of yesterday's session; **8.** Main theme: "Introduction to Participatory Monitoring and Evaluation"
10.30 am		**3.** Main theme: "Participatory Technology Development" – overview and main concepts	**5.** Main theme: "Strengthening Farmers' Experimentation"	Planning exercise in 4 groups	Exercise on participatory monitoring; reflection on the exercise
12.30 pm		LUNCH			
1.30 pm	Opening and introduction; **1.** Main theme: "Reflection on Experience in PRA"	**4.** Main theme: "Farmers' Experiments" – introduction to the subject and preparation for fieldwork	**6.** Main theme: "Selection of Options" – simulation exercise	Presentation of workgroups	Preparation for the closing session
4.00 pm	**2.** Main theme: "Problem Analysis and Problem Tree"	Fieldwork	Time to relax	Continuation of presentation	Closing session
8.00 pm		DINNER			

Source: PMHE (1992)

BOX 6
SCHEDULE OF A 4-DAY INTRODUCTORY PTD TRAINING

PARTICIPANTS: Mostly Sri Lankan researchers, with some extension workers. The training was aimed at strengthening the farmer orientation of the participants, and their skills in using participatory methods and techniques.

1994:	19 October	20 October	21 October	22 October
Session 1 08.00–09.00		Visualising linkages "research–extension–farmers"	Exercises on tools and methods as requested by the participants	Analysis and presentation of village work
Session 2 10.15–12.15		Participants' experience in farmer participation	Exercises on tools and methods (continued)	Action plan
Session 3 13.30–15.00		Farmer participation at different stages in research and extension	Preparation for village workshop	Evaluation 15.00: Departure
Session 4 15.15–17.00	Opening	Farmer participation at different stages in research and extension (continued)	Travelling to villages	
Evening Session	History of agricultural development	*Videos:* "Experience of the NWPDZ Project" & "ICRISAT"	Village work	

Source: Welligmann (1994)

BOX 7 EFFECTIVE USE OF PLENARY, INDIVIDUAL AND GROUP SESSIONS

PLENARY SESSIONS are useful:

- to make decisions about the programme
- to introduce a theme and clarify the learning objectives of a unit
- to exchange and compare information generated by sub-groups
- to add bits of theory, to generalise, and to build new guidelines for participants' behaviour
- to assess the learning process

SUB-GROUP SESSIONS are useful:

- to mobilise and analyse experiences
- to practise new skills
- to prepare the participants for change and action

INDIVIDUAL ACTIVITIES are useful:

- to prepare for group sessions (mobilisation of participants' own experiences)
- to integrate the learning into participants' existing knowledge and values
- to relate the learning to individual job and work situations

BOX 8 BASIC PRINCIPLES OF SUCCESSFUL TRAINING EVENTS

1. **START FROM WHERE THE PARTICIPANTS ARE AND STAY WITH THEM**

    Begin at the entry level of the participants and develop their understanding and skills step by step.

2. **KEEP IN MIND THE BASIC CONDITIONS FOR LEARNING such as...**
    - feeling respected and personally involved;
    - a climate of trust and mutual acceptance;
    - a climate of openness in which mistakes can be made and feelings expressed;
    - a creative climate challenging participants to explore and discover.

3. **PROVIDE FOR CONTINUOUS EVALUATION**

    Continuous (self-)evaluation is a precondition for participatory learning. Plan moments at which to evaluate with the participants "how we are doing", and vary the evaluation techniques used. Realisation of progress made and participation in directing the training strongly facilitate learning.

4. **PRESENT A GOOD STRUCTURE AND SEQUENCE**

    Participants need to have a clear idea about what they are going to do and why. A simple "structure" may include:

    a. OVERVIEW:
    what are the elements and how are they interrelated?

    b. ELEMENTS:
    dealing with the different elements one by one in a logical order.

    c. INTEGRATION:
    relating what has been learnt about the elements to the general theme and showing how the issues are linked.

5. **PROVIDE FOR VARIATION IN LEARNING ACTIVITIES**
    - alternate thinking/reasoning with experiencing/doing;
    - alternate plenary, individual and group sessions;
    - alternate between intensive and slack periods, learning and recreational periods;
    - alternate writing, feeling, seeing, listening and speaking activities of the participants;
    - use a variety of learning techniques and procedures – cases, games, exercises, drama and role play, incident technique, simulations, excursions.

6. **ENSURE CONTINUOUS VISUALISATION OF THE MAIN DISCUSSION POINTS AND FINDINGS**

    Flipcharts, cards and overhead sheets are just a few of the possible ways to do this.

7. **MAINTAIN BALANCE BETWEEN TASKS AND PROCESS**

    Give sufficient attention both to realising learning objectives and to creating an inclusive learning climate and group spirit.

8. **PROVIDE FOR FLEXIBILITY**

    The programmes should allow for adaptation to the needs and interests of the participants. The trainers may prepare possible alternative modules based on the pre-planning assessment of participants' needs and interests.

9. **ENSURE INTEGRATION**

    Relate new learning to existing knowledge and values. "Package" the learning in concepts and models that are easy to remember. Link different elements studied to main themes.

10. **PLAN FOR TRANSLATION TO THE PARTICIPANTS' OWN WORK SITUATION**

    Study the implications in participants' own situation of issues raised. Develop planning for follow-up action. Prepare for continued learning-while-doing.

BASIC ORIENTATION AND SKILLS

PART 1

PART 1 / BASIC ORIENTATION AND SKILLS

1·A	**LOOKING AT PARTICIPANTS' EXPECTATIONS AND EXPERIENCES**		25
1·B	**TOWARDS SUSTAINABLE AGRICULTURE**	■ Understanding sustainability	30
		■ LEISA, HEIA and traditional agriculture	31
		■ LEISA principles	32
		■ Changing towards LEISA	33
1·C	**A CLOSER LOOK AT FARMER PARTICIPATION**	■ Participation: what do we mean?	41
		■ Why promote farmer participation?	43
		■ Approaches to technology development (TD)	43
1·D	**PTD: FRAMEWORK AND KEY FEATURES**	■ The PTD framework	49
		■ Key features	52
		■ Roles of fieldstaff in PTD	52
		■ Obstacles to participation	53
1·E	**RESPECTING RURAL LIFE**	■ The complexity of farmers' livelihood systems	59
		■ Indigenous knowledge	60
		■ Farmers' criteria	60
		■ Community organisational structure	61
		■ Conflicting interests	61
		■ Cultural identity	62
1·F	**GENDER SENSITIVITY**	■ Gender differences	67
		■ Gender analysis	68
		■ Implications for PTD practice	69
1·G	**SKILLS IN COMMUNICATION AND PERCEPTION**	■ Dialogical communication	75
		■ Listening with an open mind	75
		■ Probing	76
		■ Body language and non-verbal communication	77
		■ Perception	77

BOXES			
	1.1	Examples of the application of LEISA principles	32
	1.2	Examples of the three-stage change towards LEISA	33
	1.3	Talking positively about PTD	57
	1.4	Gender and the use of the forest: the case of Nepal	68
	1.5	Ten areas of concern in a gender analysis within PTD	69
	1.6	Gathering information on gender roles from secondary sources	69
	1.7	Giving women farmers greater access to meetings and courses	70
	1.8	Guidelines to include a gender perspective in PRA activities	70
	1.9	Listening techniques	76
	1.10	Examples of probing questions in dialogue with a farmer	77
	1.11	Open questions to stimulate farmers' ideas	77
	1.12	Communication without words	78

TABLES			
	1.1	A typology of participation in development programmes	42
	1.2	Types of farmer participation in research	43
	1.3	Main characteristics of three approaches to technology development (TD)	44
	1.4	The PTD framework	50

HANDOUTS			
	1.1	Analysis of technology development cases: assignment for groupwork	47
	1.2	Old and new stereotypes of women and men farmers	71
	1.3	"The Bean Experiment"	72
	1.4	Measures to increase the number of female beneficiaries	74
	1.5	Interviewers' questions	81
	1.6	Farmer's comments	82

UNIT 1·A / LOOKING AT PARTICIPANTS' EXPECTATIONS AND EXPERIENCES

OVERVIEW OF THIS UNIT

EXPECTED RESULTS

Field-based development workers can be motivated to seek alternatives to their present methods of working, if they become aware that they face common problems. Such awareness may be reached through critical reflection and exchange of their own experiences. However, not all of the problems they identify can be dealt with by their own organisations, for example, through training.

After completing the learning activities in this unit, the participants are expected:

- to have developed a critical awareness of present problems in their work with farmers;
- to be able to assess these problems in terms of whether or not they can be dealt with by their own organisations;
- to have defined their own priorities for learning.

MAIN CONCEPTS

- **Learning needs:** based on the participants' experiences, the questions for which they would like to find answers during the training.
- **Participatory learning process:** training starts with the identification of the learning needs of the participants and is implemented together with them.
- **Parallel between workshop approach and PTD:** the trainer(s) and trainees collaborate in the workshop in a way similar to the interaction between farmers and development workers in the field.

TRAINING METHODOLOGY

In essence, this unit suggests that the trainer(s) and trainees carry out a participatory situation analysis, using the methods discussed in Part 2 of this book (pp. 85–140). Particular consideration should be given to those methods which may be relevant for later use with farmers. Trainers must, however, be aware that the activities in this unit take place at the very beginning of a workshop. Participants may not yet know each other well, they may not be used to free and open discussions, and they will need time to relax and to develop a group spirit. It is therefore advisable to make frequent use of the "safer" smaller groups rather than long periods in the plenary group, and to focus on concrete rather than theoretical questions. Moreover, it may only be during the training that the real learning needs become clear (just like in PTD).

LEARNING ACTIVITIES

1. Getting to know each other I
2. Getting to know each other II
3. Analysis of work experiences
4.. Formulating participants' learning agenda

DISCUSSION

Any learning situation with PTD as its central theme should put into practice the principles it tries to promote. This parallel between the participatory learning process during the workshop and the PTD process is reflected in the following:

- if farmers are supposed to play the central role in the PTD process and control its course, workshop participants should also have the feeling from the start that they determine their own learning in the workshop;
- if the PTD process starts by discussing with farmers their situation and their problems, a similar discussion with the participants is needed at the start of a workshop. This should lead to an agreement on "what to do about it", i.e. on the learning needs;
- just as a joint analysis of numerous farmers' experiences can lead to the identification of common fundamental problems, joint analysis of numerous fieldworkers' experiences can help identify strategy issues for the group as a whole.

It is a challenge for every facilitator of a participatory workshop to ensure that good use is made of the results of this situation analysis and discussion of learning needs. Usually, a large part of the workshop schedule will already have been prepared in advance. Facilitators can relate some of the identified learning needs to sessions already foreseen in the workshop and make sure that the other needs are recorded for attention when planning later workshops. During planning of the present workshop, several sessions can be kept open so that the trainer can organise learning activities directly related to needs expressed by the participants. Throughout the rest of the workshop, frequent reference can be made to these first common experiences with participatory analysis, and reflecting on this can increase the understanding of its potential and limitations.

The difficulties faced by fieldstaff are diverse, but most of them are likely to fall into one of the following categories:

- **difficult biophysical conditions:** soils, rainfall ...
- **lack of farmer interest:** low rate of adopting innovations, lack of cooperation between farmers ...
- **social situation at community level:** opposition from the powerful, conflicts between factions ...
- **lack of facilities and support from own organisation:** transport, housing, communication with head office ...
- **unsupportive external environment:** lack of co-ordination with government offices, policies for pricing, etc. ...

A subsequent analysis focuses on distinguishing problems which are "manageable" within the fieldworkers' own context (focus on community level, interaction with farmers) and other problems (national policies, organisational structure).

LEARNING ACTIVITY 1
GETTING TO KNOW EACH OTHER I

45–60 min

Objectives:
- To create a working atmosphere conducive to a participatory group learning event.
- To make participants aware of each other's background, thus creating a pool of knowledge of those attending the workshop.

Setting/approach:
- Plenary activity in which each participant writes major items of his/her background on a large sheet of paper and explains them to the others.

Materials:
- Newsprint sheets and markers – one set per participant.

Procedure:
- Distribute the newsprint and markers to the participants and ask each to write down some major items of his/her personal and working background, e.g.:
- Ask the participants to pin the sheets on the wall and ask each to present the main points briefly to the plenum.
- The newsprint sheets may be hung in one place during the workshop as a reminder of the pool of knowledge present in the group. This exercise makes the participants familiar with the value of visualising the main points of discussion for all to see.

Variations:
- This activity can be started with a brief "getting to know each other" discussion in pairs.
- With regard to their working situation, participants may also be asked to mention one or more experiences that they can contribute to the workshop, and one or more questions they wish the workshop to address.

Source: Ullrich & Krappitz (1985)

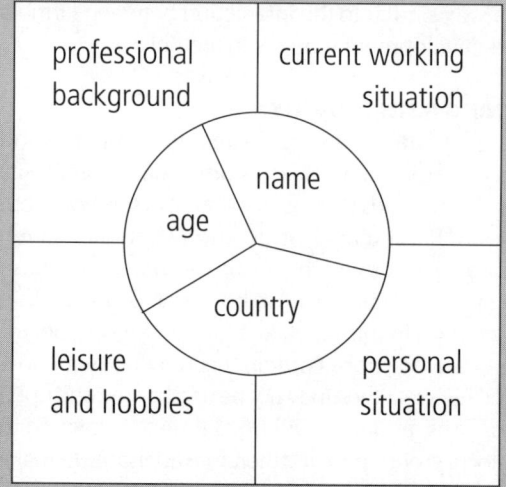

LEARNING ACTIVITY 2
GETTING TO KNOW EACH OTHER II

1–2 hrs

Objective:
- Apart from getting to know each other – to encourage creativity and inventiveness and the use of drawings to help in communication.

Setting/approach:
- A quiet meeting place large enough for the participants to work in small groups.

Materials:
- Newsprint sheets and coloured felt pens for each small group.

Procedure:
- Divide the participants into small groups of 3–5 persons.

- Ask each group to spend 20–30 minutes getting to know each other, then another 15 minutes making a collective drawing on newsprint to describe the group and its members. No words may be used in the drawing. The picture may be as simple or as elaborate as the group wishes. This is not a drawing contest. The quality of the artwork is not important.

- Ask each small group to decide how to use the drawing to introduce its members to the entire group.

- After all groups have reported, spend a few moments discussing how the drawings contributed to communication. Summarise by reviewing with participants the range of interests and abilities that were illustrated by the drawings.

What happened (at NCCK Workshop in Kenya, 1979): Some participants entered into the activity whole-heartedly; others were reticent. The pictures they produced (see the drawings on the right) needed some explanation to understand how they showed group members, but the exercise was fun and a good spirit of group cohesiveness developed.

Source: Crone & St John Hunter (1980)

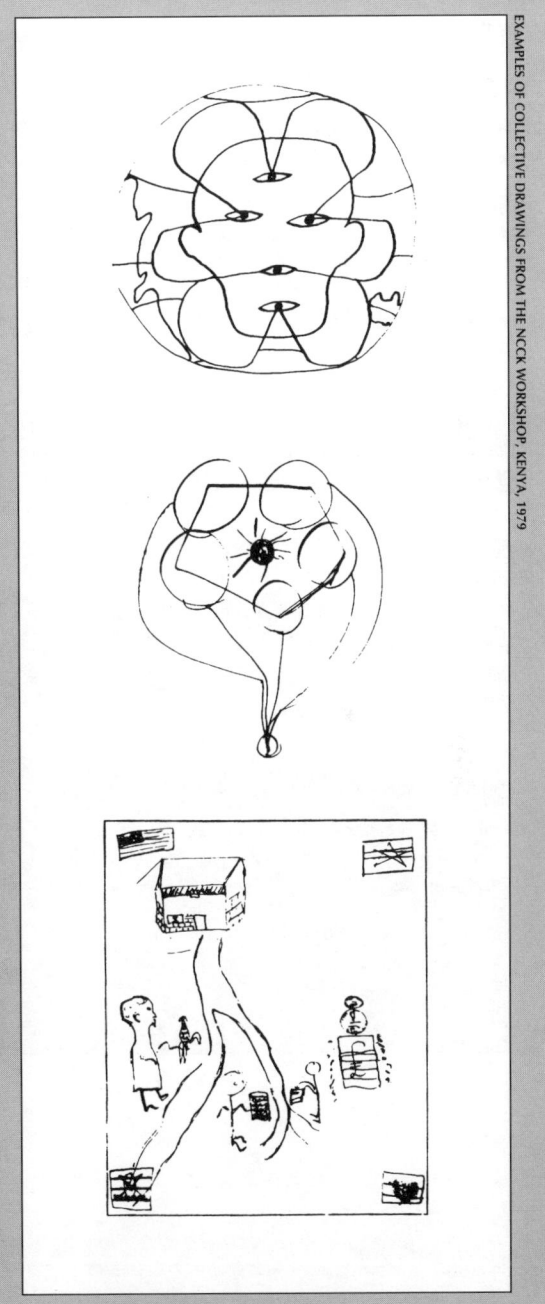

EXAMPLES OF COLLECTIVE DRAWINGS FROM THE NCCK WORKSHOP, KENYA, 1979

LEARNING ACTIVITY 3
ANALYSIS OF WORK EXPERIENCES

2–3 hrs

Objectives:
- To enhance awareness and understanding of difficulties in participants' work with farmers.
- To promote participants' commitment to deal with some of these difficulties through joint learning.

Setting/approach:
- Example of a group problem census, as discussed in Unit 2.C, pp. 115–6, combining discussion in small groups with a plenary discussion. It is especially useful in training events of several days.

Materials:
- Pen and paper, small cards and markers, board to pin the cards on.

Procedure:
- After explaining the background and objectives of this activity, outline the general procedure. Take time to explain how to use the cards, if participants are not yet familiar with using them, and point out the main advantages of using them.

- Ask the participants to form small groups of 3–5 persons and give them the following instructions:
  - discuss together the question: "What difficulties do you face when trying to work with farmers to improve their situation?";
  - take time to explain each problem thoroughly, giving examples;
  - write the essence of each problem (3–5 words at most) on a separate card;
  - hang your cards together on the board (45 minutes in total).

- Facilitate a plenary discussion of the results of the small groups:
  a. each group briefly presents their cards;
  b. clarify the cards not yet understood by other participants;
  c. structure the discussion by clustering cards with similar problems (possible categories are mentioned in the discussion section of this unit – p.26); jointly find a label for each cluster that expresses what the cards in that cluster have in common;
  d. jointly identify those problems within the "reach" of fieldworkers, i.e. that they can work on themselves, and those outside their influence that have to be accepted as "given";
  e. jointly discuss what this analysis implies for the present training.

Variations:
- Parts d. and e. of the plenary discussion may be omitted, if the previous part of the activity has consumed too much of the participants' energy; working with cards requires great concentration.

- Alternatively, the questions in Parts d. and e. may be examined first in pairs (buzzing groups) sitting next to each other.

LEARNING ACTIVITY 4
FORMULATING PARTICIPANTS' LEARNING AGENDA

45–60 min

Objectives:
- To define main areas of participants' interest for further discussion and learning during the workshop.
- To promote participants' commitment to dealing with difficulties in their work through joint learning.

Setting/approach:
- Builds on a problem analysis such as described in the previous learning activity. Participants individually define their main areas of interest.

Materials:
- Pen and paper, newsprint and markers.

Procedure:
- Ask participants, on the basis of the results of the previous discussion about difficulties in their work, to list issues they would especially like to address during this workshop or perhaps subsequent ones.
- Suggest that each participant write his/her list in large letters on newsprint and hang these on the wall for brief plenary reflection.
- Comment briefly on the different lists and suggest possibilities as to when and how different issues could be discussed; if some issues do not find a place in the present workshop, be open about this.

Variations:
- Participants may be asked to interview each other to generate main areas of interest. This provides an opportunity to practise helping others to formulate problems or express areas of concern.
- If participants are asked to write their areas of interest on cards (one issue per card in large letters), similar suggestions from different participants can be grouped together to create a joint agenda for the group. The clustering would take another 30 minutes, at least.
- A similar activity can be repeated at other times during the workshop – for example: after an overview has been given of the PTD process in general (see Unit 1.D, pp. 49–58), to find out which parts are of special relevance for the participants; or at the end of the workshop, to generate ideas for subsequent ones.

UNIT 1·B / TOWARDS SUSTAINABLE AGRICULTURE

OVERVIEW OF THIS UNIT

EXPECTED RESULTS

This unit helps participants to grasp the essential concepts of sustainable agriculture and relate these to the situation in their own working area. After completing the learning activities in this unit, the participants are expected:

- to understand various aspects of the concept of sustainability;
- to be able to distinguish critically between traditional agriculture, high-external-input agriculture (HEIA) and low-external-input and sustainable agriculture (LEISA);
- to understand the main principles of LEISA as a basis for choosing suitable LEISA practices and adapting these in the local situation;
- to understand the importance of step-by-step change towards LEISA and be able to define the role that PTD plays in this process.

MAIN CONCEPTS

- **Sustainability:** Resource management which satisfies human needs while maintaining the quality of the environment and conserving natural resources.
- **Traditional agriculture:** Forms of agriculture that are based on indigenous knowledge and have evolved over many generations.
- **High-External-Input Agriculture (HEIA):** Forms of agriculture which depend on significant levels of inputs, such as fertilisers, pesticides and fossil energy, from outside the farm or farming area.
- **Low-External-Input and Sustainable Agriculture (LEISA):** Forms of agriculture based mainly on inputs coming from the individual farm or farming area, in which deliberate action is taken to ensure sustainability.
- **Mimicking nature:** Effort made in LEISA to imitate the nutrient accumulation mechanisms found in natural ecosystems.
- **Seeking diversity:** Striving for different types of crops, animals and enterprises within one farm or region, and for supportive relations between these to increase stability.
- **Living soil:** The quality of soil and the life processes in it, which form the basis of any agricultural activity.
- **Cyclic flow patterns:** Preventing unnecessary losses of nutrients, water and energy by means of recycling and compensating for exports and sales.
- **Transition:** Gradual change towards sustainable agriculture.
- **Efficiency, adaptation and redesign:** The increasingly complex and further-reaching stages in transition to sustainable agriculture.

TRAINING METHODOLOGY

The approach begins with critical reflection on past and current developments in the participants' own area. Checklists of questions can help to focus this analysis. Examples of alternative approaches to agricultural development can be shown using audiovisual aids (see Appendix, pp. 222–3). The discussion section of this unit can be distributed to support the analysis made by the participants.

LEARNING ACTIVITIES

1. The Nuts Game: experiencing sustainability
2. Characterising the existing farming system
3. Clarifying terminology
4. Experiencing cyclic flow patterns
5. Farmer participation for LEISA: the game of opposites (follow Learning Activity 1 of Unit 1.C, p. 45)

DISCUSSION

UNDERSTANDING SUSTAINABILITY

Globally, agriculture is under pressure to maintain a growing population. In more and more areas, the attempt to raise agricultural production is leading to overuse of the natural resources, reflected in loss of genetic diversity, deforestation, erosion, etc. These forms of environmental degradation cause long-term decline in agricultural productivity, as they deplete the resource base itself. There is a need for sustainable forms of agriculture which satisfy human needs while maintaining the quality of the environment and conserving natural resources. Sustainable agriculture is:

- **economically viable:** farmers produce at an adequate and stable level, and at a risk level which is acceptable to them;
- **ecologically sound:** the quality of the environment is maintained or enhanced and natural resources are

conserved. Ecologically sound agricultural systems are healthy and highly resistant to stress and shock;

- **socially just:** the agricultural system assures equal access to land, capital, information and markets for all people involved, whatever their socio-economic position, sex, religion or ethnic group;

- **humane:** all forms of life (plant, animal, human) are respected and treated with dignity;

- **adaptable:** sustainable rural communities are able to adjust to constantly changing conditions such as population growth, and new policies and market demand (Gips, 1986).

LEISA, HEIA AND TRADITIONAL AGRICULTURE

In order to gain a deeper understanding of the current situation of farming in a given area, it is helpful to distinguish between three approaches to farming and their major characteristics:

- **Traditional agriculture** is based on indigenous knowledge and practices that have evolved over many generations. It is generally oriented to subsistence, uses resources available locally and makes little use of external inputs. Traditional agriculture is highly varied, as it depends on site-specific ecological and cultural factors.

Confronted with rapid changes such as increasing population pressure and greater need for cash, farmers practising traditional agriculture cannot always increase productivity sufficiently. They may therefore expand farming into marginal areas, which increases the risks of over-exploitation, erosion and other forms of environmental degradation.

- **High-External-Input Agriculture (HEIA)**, the conventional "modern" approach to agricultural development, puts great emphasis on the use of external inputs, such as hybrid seed, fertiliser, biocides, mechanisation and credit, to enhance productivity. HEIA is characterised as follows:

  - it uses high levels of external inputs;
  - it involves strong links between farmers and commercial and governmental services;
  - it is market-oriented;
  - it is specialised in only a few crops grown in pure stands or single-purpose livestock kept in large numbers;
  - the biomass in the landscape is greatly reduced.

HEIA has certain advantages, such as short-term increase in production and cash income, uniform production processes and lower labour costs. However, it also has many disadvantages:

- it has limited applicability to dry and risk-prone farming areas;
- it has negative impacts on water, air and human health;
- it tends to erode soils, genetic resources and local knowledge;
- it cannot be applied by many poor farmers and in poor areas;
- it under-utilises resources available locally;
- it over-utilises non-renewable resources such as fossil energy and phosphorus;
- it increases the dependency of farmers.

These disadvantages, combined with recent increases in costs of external inputs as a result of "structural adjustment", have stimulated interest in developing sustainable farming practices. New approaches have emerged under various labels, e.g. Biodynamic, Ecological, Natural, Organic, Permaculture, and Regenerative Agriculture. We use the term "LEISA" to refer to them all.

- **Low-External-Input and Sustainable Agriculture (LEISA)** depends primarily on resources from the farm, village and region, and is characterised as follows:

  - it aims to integrate soil fertility management, arable farming and animal husbandry;
  - it makes efficient use of nutrients, water and energy and recycles them as much as possible, thus preventing depletion and pollution;
  - it uses external inputs only to compensate for local deficiencies;
  - it involves site-specific farming practices;
  - it incorporates the best of indigenous knowledge and practices, sustainable agricultural experiences and conventional scientific knowledge;
  - it aims at stable and long-lasting production levels (Reijntjes et al., 1992).

LEISA PRINCIPLES

There is no fixed set of LEISA technologies. These need to be developed or adapted according to the specific agro-ecological situation and the needs of the farm household. To be able do this, a good understanding of the basic principles of LEISA is needed:

■ **Mimicking nature:** All natural ecosystems without human disturbance manage to accumulate nutrients against the forces of erosion, runoff, fire, leaching and volatilisation. In tropical ecosystems, nutrient accumulation is based on five mechanisms:

- living plants form a continuous soil cover;
- litter layer of decomposing leaves covers the soil;
- the major period of nutrient release by microbes coincides with the major nutrient demand period of plants;
- most nutrients are retained in living plants or animals;
- roots of different plants are distributed throughout the soil at different depths (Woodmansee, 1984).

Our ability to develop sustainable farming systems depends largely on how successfully we can include these mechanisms in our agricultural practices.

BOX 1-1 EXAMPLES OF THE APPLICATION OF LEISA PRINCIPLES

Mimicking nature	■ multi-storey agroforestry systems ■ mulching
Seeking diversity	■ mixed cropping of cereals and legumes ■ crop–livestock integration ■ mixtures of different varieties or breeds ■ cereal–fish culture ■ multiple sources of nutrients
Living soil	■ mulching, cover crops ■ contour bunding, windbreaks ■ organic matter to feed soil life ■ use of botanical pesticides
Cyclic flow patterns	■ use of crop residues as fodder ■ composting of kitchen waste ■ deep-rooting crops

■ **Seeking diversity:** Natural ecosystems consist of many different plant and animal species interacting with each other. The resulting elaborate web of inter-relations gives strength to the ecosystem, enabling it to resist disturbances such as erratic rainfall and attacks of pests and diseases. At farm level, diversification of species, varieties, breeds and enterprises decreases vulnerability to external disturbances, not only climatic but also economic. Growing diverse species also permits better use of varied environments (e.g. a field with differences in soil fertility) and allows beneficial combinations to be made. In ecosystems, these beneficial combinations develop naturally over a long time, but in agro-ecosystems the farmers create the combinations themselves.

■ **Living soil:** One of the most important components of the soil is soil life, including bacteria, fungi, algae, protozoa, nematodes, beetles, centipedes, termites and earthworms. Soil life plays a major role in many essential processes which determine nutrient availability and recycling and, thus, agricultural productivity. Farmers have to create favourable conditions for soil life and manage organic matter so as to create a fertile soil in which healthy plants can develop. Soil can be protected by vegetative cover to decrease rain impact and heating by the sun, and by mechanical measures to limit erosion by water and wind. Soil life also needs to be protected from harmful man-made substances such as pesticides and fungicides. Organic matter must be provided to feed soil life.

■ **Cyclic flow patterns:** In a natural ecosystem, hardly anything is lost; nutrient and other cycles are almost closed. LEISA aims at learning from these natural recycling processes to prevent depletion of natural resources. Losses are minimised through cover crops, deep-rooting species that recycle nutrients leached from the topsoil, erosion control, and improved collection, storage and application of wastes from crops (residues), livestock (manure and urine) and the kitchen (food wastes). Nutrients that are "exported" in crop and animal products are replaced by symbiotic nitrogen fixation, organic matter from elsewhere, complementary use of fertilisers and feed supplements. Similarly, water flows are managed so that optimal use is made of the available water.

CHANGING TOWARDS LEISA

Changing from current farming practices to more sustainable practices is normally a gradual process of transition. A phased approach to transition which starts with the easier changes helps minimise risks and spread investments. It allows farmers to develop the required skills gradually and to gain self-confidence before tackling the more complex changes. In this transition process, the following phases can often be distinguished (MacRae et al., 1990):

- **Increased efficiency:** Current practices are altered to reduce both consumption of resources and environmental impact. Lower losses and higher effectiveness make it possible to use fewer inputs but still obtain the same level of production. Innovations introduced at this stage do not require substantial changes in agricultural practices.

- **Adaptation:** Substantial adaptation of agricultural technologies and techniques occurs. Finite resources and environmentally disruptive techniques are replaced by more environmentally benign methods. Natural processes such as recycling of nutrients are consistently integrated.

- **Redesign:** A structured change is made of the entire farming system to ensure full inclusion of the principles of sustainable agriculture, by mimicking the characteristics of natural ecosystems. This phase is more complex and considerably longer than the previous two. Changes are more fundamental and require time to realise their full impact (e.g. development of a balanced soil life).

In the course of these three phases, the interventions increase in size, impact and complexity. However, farmers rarely follow these phases in a strictly linear way. A first change may be made in Efficiency, but certain aspects of Adaptation and Redesign may follow soon and be implemented simultaneously. Fine-tuning and re-adjustment of farming technologies may involve some repetition of the phases (iteration).

Seeking sustainability in agriculture requires careful integration of new scientific insights with local knowledge and practices. This is the essence of PTD. Thus, LEISA and PTD are fundamentally two aspects of the same approach. On the one hand, pursuing PTD without a clear LEISA perspective may lead to less sustainable farming practices. For example, farmers may opt for trying out higher pesticide doses or mixing pesticides with kerosene to make them more effective or cheaper. On the other hand, pursuing LEISA without PTD is impossible. The farmers themselves play the key role in changing towards LEISA because:

- changing to LEISA starts from a thorough understanding of the current situation, actual problems and their context; and only the farmers themselves have an intimate knowledge of these;

- solutions adapted to site-specific conditions, rather than blanket recommendations, need to be developed; and the formal research and extension systems lack the capacity to develop these multiple adaptations;

- in transition from HEIA to LEISA, excessive use of external inputs is "replaced" by a much stronger role of farmers in managing the various resources; and close collaboration between development agents and farmers can help increase their insight and management capacities;

- sustainable agriculture must be maintained in vulnerable ecosystems under ever-changing economic conditions, and must therefore be constantly monitored by the farmers themselves.

BOX 1-2 EXAMPLES OF THE THREE-STAGE CHANGE TOWARDS LEISA

Increased efficiency	■ improved crop combinations to ensure more efficient use of sunshine and rainfall
	■ monitoring pests and intervening only when the economic damage exceeds intervention costs (Integrated Pest Management)
Adaptation	■ planting along contours instead of down the slope
	■ crop–livestock integration
	■ biological measures of controlling pests and diseases replace use of chemicals
Redesign	■ new crop rotations to increase soil fertility
	■ preventive pest-management practices implying changes in, for example, cropping systems, tillage, fertilisation or sowing

LEARNING ACTIVITY 1
THE NUTS GAME: EXPERIENCING SUSTAINABILITY
50 min

Objective:
- To allow participants to experience the concept of sustainability.

Setting/approach:
- Repeated activity of a small group of participants in a plenary session, while the other participants observe.

Materials:
- Poster with goal and rules of the game (see p.35), recording forms (see format given below), an open bowl (about 30 cm diameter) and about 140 nuts, pebbles or seeds of 1–2 cm diameter.

Procedure:
- A small group (4–5 people) sits around an open bowl containing 25 nuts. The trainer introduces the exercise by unveiling the written goal and rules of the game, which are read by the participants in silence. When the trainer gives the signal, the game starts. When the game is completed, the total harvest per person and the group total are recorded (10 minutes).
- After one group has finished the game, 4–5 new persons are invited to play it. This can be repeated three times or more (20 minutes).
- After the games, the following two questions are discussed in the plenum:
  - How did you feel during the game?
  - What did you learn during the game?

Important items for discussion are cooperation, self-restraint, trust, the regenerative capacity of natural resources, depletion, total harvest and equity in division of the harvest (20 minutes).

In this game, the bowl symbolises the resource pool (e.g. the soil), the nuts represent the resources themselves (e.g. the crops or soil nutrients harvested) and the replenishment cycle symbolises natural rates of resource regeneration.

Note:
The first game usually ends after a few seconds because, when the starting signal is given, all the participants simultaneously grab all the nuts they can get. Thus, they empty the bowl and no refilling takes place. The second and third games usually take a bit longer and are played for several rounds. In case this does not happen, the trainer might stimulate a brief discussion about similarities between the game and real-life situations, directing the discussion towards such concepts as sustainability, non-exploitative use of resources and Nature's regenerative capacity. After this intermission, a new group of players should be able to play the game for several rounds, thus achieving a higher total score per player and as a group.

Source: Edney (1979)

REPORTING FORMAT

	Game 1	Game 2	Game 3	Game 4
Player 1 Player 2 Player 3 Player 4 Player 5				
Total				
Minimum/Maximum				
Average				

POSTER

THE NUTS GAME

GOAL: ● Each player's goal is to get as many nuts as possible during the game.

RULES: ● Upon the organiser's signal, the players take out nuts from the bowl – all at the same time, but using only *one* hand. This makes one "round".

● The balance left in the bowl is doubled after each round by the organiser, up to the maximum of 25 nuts.

● The game is over when the bowl is empty, or after 10 rounds.

LEARNING ACTIVITY 2
CHARACTERISING THE EXISTING FARMING SYSTEM

60–80 min

Objectives:
- To become aware of the main characteristics of the existing farming system in the participants' working area, and of its advantages and disadvantages.

- To become aware of the main characteristics, advantages and disadvantages of traditional agriculture, HEIA and LEISA.

Setting/approach:
- Discussion in small groups, followed by presentation and discussion of the groups' findings in the plenum.

Materials:
- Blackboard, pens and paper; or newsprint, cards and markers.

Procedure:
- The trainer introduces the subject and explains the objectives of the learning activity (5–10 minutes).

- Two or three participants working in a similar area are grouped together and requested to "buzz" (discuss quietly with each other) about the following questions (20 minutes):

  • How would you characterise farming in your area, and to what extent does this resemble traditional agriculture, HEIA and/or LEISA as outlined in the discussion section of this chapter?

  • What are the strong and weak points of this farming system? (Think of environmental, social, cultural and economic aspects – both farm and national economics.)

- In a plenary session, the answers to the first question are listed and written on the blackboard by one of the participants. Similar answers are put in the same cluster. First, each group is asked to mention only two characteristics. After all groups have had a chance to contribute, additional answers can be given by each group. The answers that arise are likely to generate a number of variables that can be used to characterise and compare traditional agriculture, HEIA and LEISA. These variables may then be listed in a table similar to the one shown on p. 37.

- Using the list of variables thus created in the plenary session, the existing farming system(s) in the participants' areas are characterised and compared with the three archetypes.

- In a plenary discussion, the advantages and disadvantages of the existing farming system(s) are identified, using the answers of the "buzz groups" to the second question.

- The learning activity may be rounded off by asking one of the participants to summarise his/her main learning points and to mention remaining questions. Indicate where and how such questions will be dealt with in the remainder of the training session.

Variations:
- The following questions could be used in addition to or to replace the buzz group questions given above:

  • How would you characterise the type of farming promoted by the extension services and to what extent does this resemble traditional agriculture, HEIA and LEISA as outlined in the discussion section of this chapter?

  • What are the main advantages and disadvantages of this type of farming? (Think of environmental, social, cultural and economic aspects – both farm and national economics.)

- Referring to the same variables indicated above, the main characteristics of the farming system promoted by the extension service (in most cases, strongly oriented to HEIA) can be discussed, and a need for alternative farming systems can be considered.

Source: van der Werf (1996)

SAMPLE TABLE

Variables	Traditional	HEIA	LEISA
Variety / specialisation			
Use of locally available inputs			
Use of external inputs (fertilisers, pesticides, etc)			
Use of local knowledge			
Use of extension services			
Main production objectives			
Cash income			
Labour requirements			
Level of production			
Health effects			
Degree of recycling			
Degree of sustainability			
…………			
…………			
…………			
…………			
…………			
…………			
…………			
…………			

LEARNING ACTIVITY 3
CLARIFYING TERMINOLOGY

1½ hrs

Objective:
- To gain a shared understanding of several terms often used when discussing LEISA.

Setting/approach:
- Small group activity concluded by a plenary session.

Materials:
- Blackboard, pens and paper; or newsprint, cards and markers.

Procedure:
- Main terms related to LEISA are written on the blackboard, e.g. Ecology, Holistic approach, Independence, Integrated, Site-oriented, Stability, Sustainable, Synergy. Small groups (4–6 persons) are then assigned to discuss four terms each (45 minutes).
- In a plenary session, the groups present their summarised discussions in key words describing each term (45 minutes). Compare trainer's notes below.

Variations:
- Instead of dividing into small groups, the participants can discuss the terms in a plenary session, in which they start by giving key words related to each term in turn. This takes about 1 hour.
- There are many different forms of LEISA. In some training situations, it may be useful to discuss the differences between LEISA systems such as Bio-dynamic, Ecological, Indigenous, Natural, Organic, Permaculture, Regenerative and Resource-efficient.

Sources: adapted from Reijntjes et al. (1992) & van der Werf (1989)

TRAINER'S NOTES

CONCEPTUAL TERMINOLOGY

Ecology: The science of relationships between organisms and their environment.

Holistic: An approach considering all components and aspects of a system and their interrelations.

Independence: Not depending on, controlled by or relying on others/outsiders.

Integrated: Considering all components and combining them into a consistent unity.

Site-oriented: Adapted to the specific limitations and possibilities of the given area.

Stability: The degree to which productivity is constant in the face of disturbances caused by the normal fluctuation of climate and other variables.

Synergy: The action of two or more organisms to achieve an effect that is beneficial to each organism.

LEISA FARMING SYSTEMS

Bio-dynamic: Seeks to connect nature with cosmic forces and to create an integrated farming system in harmony with its habitat, avoiding use of synthetic fertilisers and pesticides.

Ecological: General term indicating consideration of ecological laws in agriculture.

Indigenous: Generated locally or elsewhere but transformed by local people and incorporated into the local way of life; may refer to knowledge, technology, etc.

Natural: Seeks to follow nature by minimising human interference – no mechanical cultivation, no synthetic fertilisers or prepared compost, no weeding by tillage or herbicides; no dependence on chemicals.

Organic: Encourages healthy soil and crops through nutrient recycling in organic matter, crop rotation, appropriate tillage, and avoiding synthetic fertilisers and pesticides.

Permaculture: A consciously designed, integrated system of perennial or self-perpetuating species of crops, animals and structures, aimed at permanent self-sustaining agriculture.

Regenerative: Stresses the idea that agricultural technology should strengthen the natural processes on which it depends.

Resource-efficient: Aimed at efficient use of resources (material, energy and human) to ensure that agriculture will be sustainable.

Sustainable: Managing resources to satisfy changing human needs while maintaining or enhancing the quality of the environment and conserving natural resources.

Traditional: Based on indigenous knowledge and practices, and evolved over time without planned external interventions.

LEARNING ACTIVITY 4
EXPERIENCING CYCLIC FLOW PATTERNS

50 min

Objectives:
- To experience the cyclic aspect of nutrient flow patterns.
- To clarify the differences in nutrient flow patterns in a natural ecosystem, in traditional agriculture and in market-oriented farming.

Setting/approach:
- Games in small groups followed by plenary discussion of group experiences.

Materials:
- Three dice (one per group), a figure to represent each participant on the nutrient flow scheme (a bean, a pebble or anything similar in size), enlarged prints of the nutrient flow schemes illustrated here and continued on p. 40. In the plenary discussion, overhead sheets of the three schemes may also be useful.

Procedure:
- After a general introduction of the LEISA principles, and particularly the cyclic flow patterns, the trainer divides the participants into three groups and introduces the game as follows:

  • In each of the 3 groups, each player takes the part of a mineral and starts in the soil. Players take turns to throw the dice. The number thrown indicates the path the player is to take on the nutrient flow scheme, e.g. from the soil into the natural ecosystem, washed down to the subsoil, taken up in grasses or shrubs, etc.

The first player to become a lion or a family member wins the game, provided he or she can explain in the subgroup what has happened in each step of the game (30 minutes).

  • In the plenary discussion, each of the groups presents its findings, and similarities and differences are discussed (20 minutes).

Source: van Noordwijk (1984)

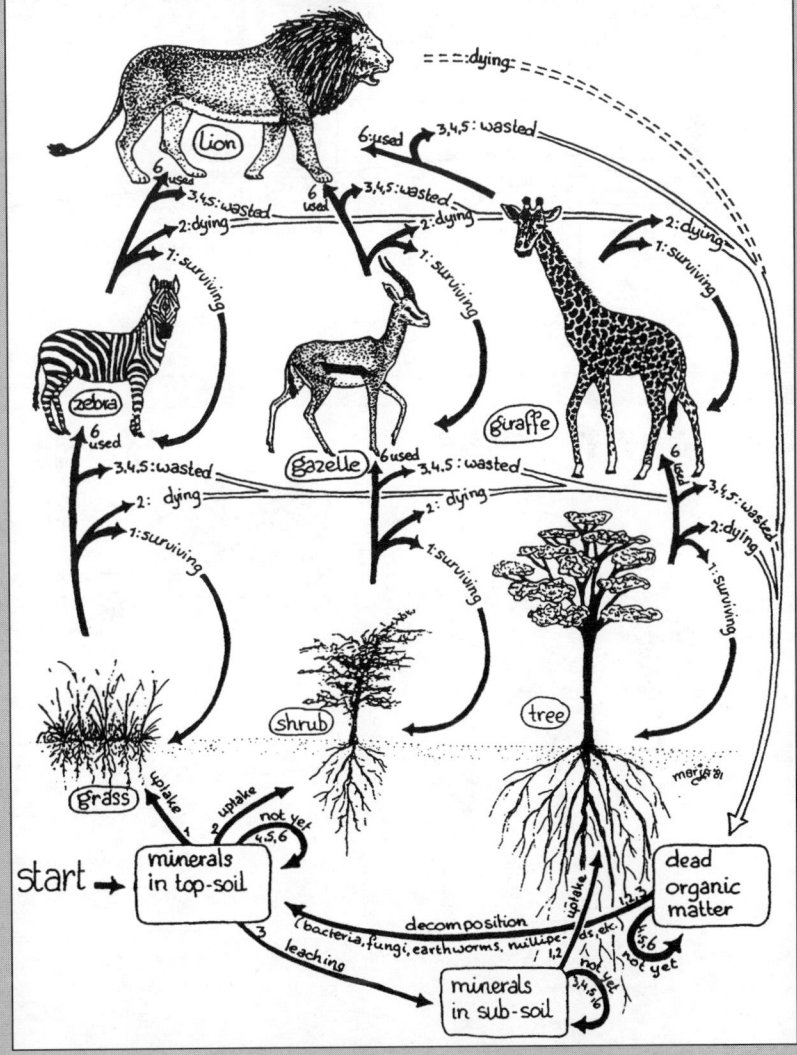

1 / Basic Orientation and Skills • B / Towards Sustainable Agriculture 39

LEARNING ACTIVITY 4 (continued)

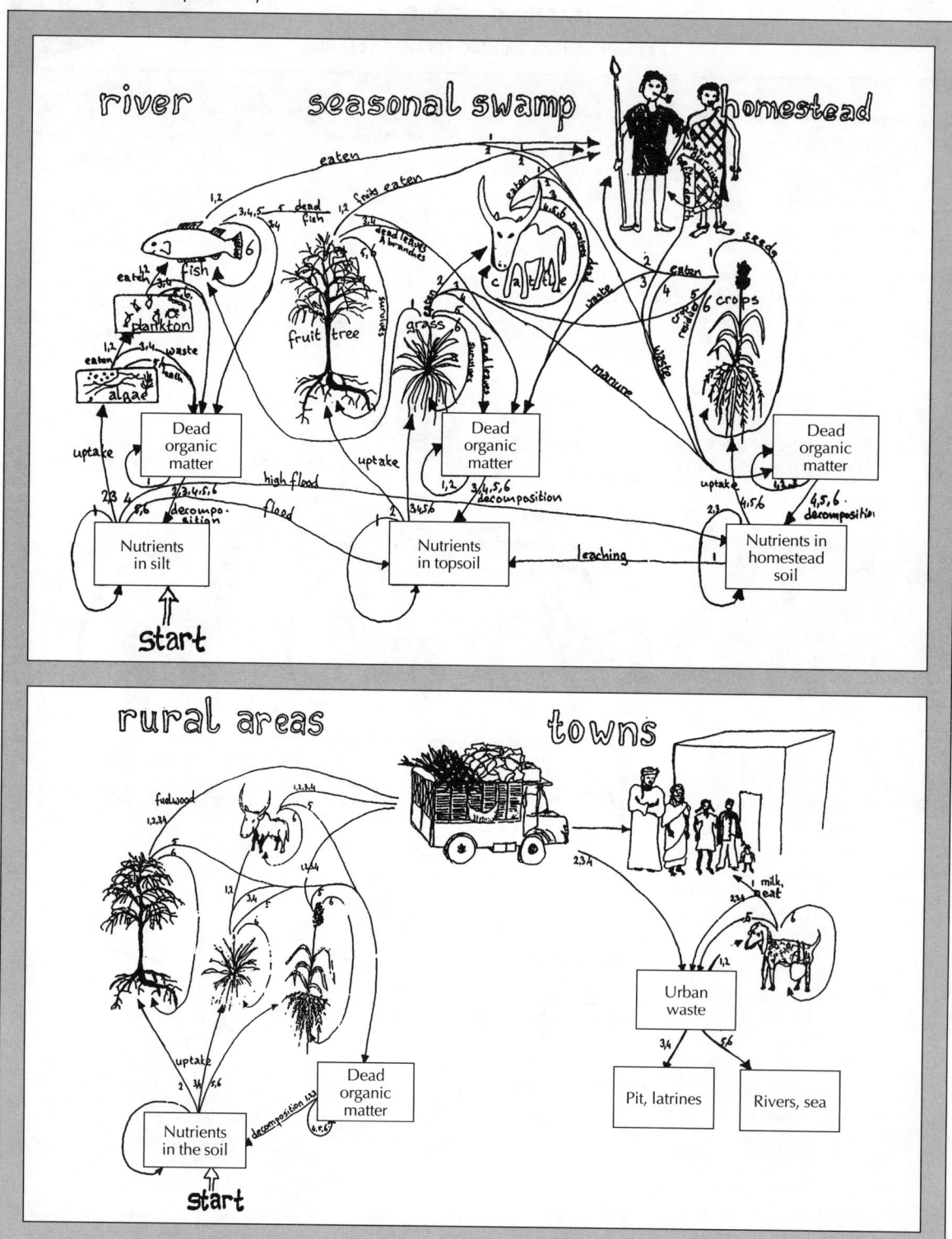

40 DEVELOPING TECHNOLOGY WITH FARMERS: A TRAINER'S GUIDE FOR PARTICIPATORY LEARNING

UNIT 1·C / A CLOSER LOOK AT FARMER PARTICIPATION

OVERVIEW OF THIS UNIT

EXPECTED RESULTS

The term "farmer participation" is frequently used to mean very different things. To be able to give this concept a proper place in their work, fieldworkers need to know the varied meanings of the term, and why participation is being promoted. They should become aware of the advantages and disadvantages of a participatory approach as compared to farmers' own technology development and the conventional transfer-of-technology model.

After completing the learning activities in this unit, participants are expected:

- to recognise the important role of farmers in generating and communicating innovations, i.e. in the technology development process;
- to be familiar with the different possible levels of farmer participation and able to choose a level appropriate for their own situation;
- to understand the potential and limitations of a participatory approach to technology development as compared to farmers' own and the conventional top-down approach;
- to be able to assess critically their own (organisation's) approach in technology development.

MAIN CONCEPTS

- **Reasons for promoting farmer participation:** Pragmatic (to increase efficiency of one's work), ethical (to promote equity and right of self-determination) and political (to empower the poor).
- **Degrees of farmer participation:** From mere involvement in implementation, through consultation, to really influencing decision-making.
- **Indigenous technology development:** Innovations in agriculture which take place without interference from outsiders.
- **Transfer-of-technology (ToT):** Innovations that have been developed at research stations are transferred via extensionists to farmers – the "end users".
- **Participatory Technology Development (PTD):** Formal research and extension agencies complement and support farmers' own technology development efforts.

TRAINING METHODOLOGY

An understanding and acceptance of the issues dealt with in this unit can best be developed by studying practical experiences of technology development, rather than discussing the models on a theoretical level. Good cases can often be obtained from the participants, or they may be collected beforehand by the facilitators. Field visits to such examples will enrich the learning experience. Referring to a number of cases should stimulate participants to think of various possible options, rather than models to be copied.

In case the participants do not have relevant experiences, prepared videos or slide series can be used as basis for analysis and discussion (see Appendix, pp. 222–3). In many cases, participants' existing ideas about "farmer participation" may need to be confronted to bring about a change in attitude. Variations on the game of opposites (Learning Activity 1, p.45) may help to do this.

LEARNING ACTIVITIES

1. The game of opposites
2. Case studies in technology development (TD)
3. PTD versus ToT in West Africa
4. The IPRA method: a video

DISCUSSION

PARTICIPATION: WHAT DO WE MEAN?

"Farmer participation" is one of the most frequently used, and misused, concepts in development rhetoric in the past decade.

The term is used, for example, to refer to farmers paying for irrigation facilities, but also to farmers exerting decisive influence on the activities of extension and research organisations. In any discussion about farmer participation, its meaning in the specific context must therefore be clarified.

In agricultural extension, the categories shown in Table 1.1 (p.42) are helpful in analysing farmer participation.

In agricultural research, farmer participation generally takes one of the forms shown in Table 1.2 (p.43).

PTD is an effort to put into practice the furthest-reaching of the options given in Tables 1.1 and 1.2. In PTD,

TABLE 1·1
A TYPOLOGY OF PARTICIPATION IN DEVELOPMENT PROGRAMMES

TYPOLOGY	COMPONENTS OF EACH TYPE
1. Passive participation	People participate by being told what is going to happen, or has already happened. It is a unilateral announcement by an administration or project management, without listening to people's responses. The information being shared belongs only to external professionals.
2. Participation in information-giving	People participate by answering questions posed by extractive researchers using questionnaire surveys or similar approaches. People do not have the opportunity to influence proceedings, as the findings of the research are neither shared nor checked for accuracy.
3. Participation by consultation	People participate by being consulted, and external agents listen to views. These external agents define both problems and solutions, and may modify these in the light of people's responses. Such a consultative process does not concede any share in decision-making, and professionals are under no obligation to take on board people's views.
4. Participation for material incentives	People participate by providing resources, for example labour, in return for food, cash or other material incentives. Much on-farm research falls into this category, as farmers provide the fields but are not involved in the experimentation or the process of learning. It is very common to see this called participation, yet people have no stake in prolonging activities when the incentives end.
5. Functional participation	People participate by forming groups to meet predetermined objectives related to the project, which can involve the development or promotion of externally initiated social organisation. Such involvement usually occurs not at early stages of project cycles or planning but after major decisions have been made. These institutions tend to be dependent on external initiators and facilitators, but may become self-dependent.
6. Interactive participation	People participate in joint analysis, which leads to action plans and the formation of new local institutions or the strengthening of existing ones. It tends to involve interdisciplinary methodologies that seek multiple perspectives and make use of systematic and structured learning processes. These groups take control over local decisions, and so people have a stake in maintaining structures or practices.
7. Self-mobilisation	People participate by taking initiatives independent of external institutions to change systems. They develop contacts with external institutions for resources and technical advice they need, but retain control over how resources are used. Such self-initiated mobilisation and collective action may or may not challenge existing inequitable distributions of wealth and power.

Source: Pretty (1994)

TABLE 1·2 TYPES OF FARMER PARTICIPATION IN RESEARCH	
MODE	**OBJECTIVE**
Contractual	Scientists contract with farmers to provide land or service
Consultative	Scientists consult farmers about their problems and then develop solutions
Collaborative	Scientists and farmers collaborate as partners in the research process
Collegial	Scientists work to strengthen farmers' informal research and development systems in rural areas

Source: Biggs (1989)

"farmer participation" implies an *acceptance that local people can, to a large extent, identify and modify their own solutions to suit their needs*. It means that "outsiders" such as researchers and development agents support farmers in their own efforts to change their farming systems. This support focuses on enhancing farmers' capacity to innovate, to experiment, to develop their farming system in a sustainable way and to increase their control over resources and decision-making affecting their farms.

WHY PROMOTE FARMER PARTICIPATION?

Strong farmer participation in agricultural development is essential if sustainability is to be achieved. As already argued at the end of the previous unit, farmer participation is needed:

- to link technology development with farmers' intimate knowledge of the local situation;

- because formal research and development institutes have limited capacity to develop a multitude of locally-specific technology adaptations;

- so that indiscriminate use of external inputs can be "replaced" by farmers' day-to-day observation and decision-making about the use of inputs.

Farmer participation is being increasingly promoted, not only in connection with LEISA development. It contributes to higher rates of adoption of technologies developed by researchers, especially in resource-poor areas with highly diverse farming systems, and reduces costs of the research and extension by increasing farmers' financial contributions.

NGOs often see participation as an end in itself, in order to generate countervailing power at the grassroots level. Without such empowerment of farmers, the benefits of development are not expected to reach the grass roots.

Why do *you* advocate farmer participation? Is it for reasons of:

- effectiveness of your work? to increase rates of adoption? to achieve sustainability of agriculture?

- efficiency of your work? to reach more farmers with limited staff? to reduce costs? to increase farmers' financial contributions?

- equity and ethics? to ensure that people, especially the poor, have a say in activities that affect their lives?

- empowerment? to strengthen farmers' bargaining power against governments and private interests, so that lasting development can be achieved?

APPROACHES TO TECHNOLOGY DEVELOPMENT (TD)

A great number of innovations in farming have occurred without intervention from outside. Braidwood (1967) refers to the "atmosphere of experimentation" which characterised even the Neolithic farmer since the earliest stages of agriculture. One may call this *"indigenous technology development"*.

In modern times, institutes have been created which specialise in parts of the agricultural development process, such as research and extension. They operate according to the following model: innovations are developed at research institutes and transferred through the extension service for adoption by farmers. The term *"transfer-of-technology"* (ToT) is often used to refer to this linear model of technology development.

As a reaction to major problems encountered with ToT (cf. Merrill-Sands, 1986), participatory approaches to technology development have been developed. *"Participatory technology development"* stresses the importance of farmers' role in agricultural innovation and change, which is complemented by formal

TABLE 1·3
MAIN CHARACTERISTICS OF THREE APPROACHES TO TECHNOLOGY DEVELOPMENT (TD)

CRITERIA	Indigenous TD	ToT	Participatory TD
Objectives	secure living, reduce risks	maximise yield	farmers' agricultural self-management
Source of innovations	farmers	research organisations	farmers complemented by research organisations
Nature of knowledge	holistic	particularistic	creative tension between holistic & particularistic
Experimental approach	largely unknown	scientific procedures	farmers' methods complemented by simple scientific procedures
Channels of communication	farmer-to-farmer	extension service	*multiple system:* farmers, NGOs, extensionists, etc
Process of communication	informal, horizontal	formal, vertical, top-down	semi-formal
Role of farmers	generator of knowledge, communicator, user	receiver, adopter	generator, communicator, evaluator of outside ideas, user
Role of fieldstaff	none	teacher, control compliance with regulations	*multiple:* moderator, resource person, co-researcher, teacher

research. The main characteristics of these three approaches to technology development are summarised in Table 1.3.

In indigenous technology development, the farmers control what happens on their farm. All decisions – for example, those about which aspect of their farm system needs to be improved or which new options should be tested – are in their hands alone. In ToT, many of these decisions are taken from them; in some cases of large irrigation schemes, the farmers are little more than labourers on their own land. The PTD approach aims at giving the decision-making role back to the farmers. Where outsiders contribute to farmers' decision-making, this is done openly as equals in dialogue.

LEARNING ACTIVITY 1
THE GAME OF OPPOSITES

about 2 hrs

Objectives:
- To enhance participants' understanding of the advantages and disadvantages of adopting a participatory approach as compared with the ToT approach and/or indigenous technology development.
- To increase participants' skills in creatively "managing" conflicts on concepts, views and approaches.

Setting/approach:
- Game in a workshop setting; participants are asked to take sides in a discussion and defend that side's view in a constructive way.

Materials:
- Pen and paper, board (e.g. blackboard and chalk) to visualise main points of discussion.

Procedure:
- Present the main theme of the debate; in the context of this unit, this may be:
  - the advantages of conventional ToT versus PTD;
  - same question but limiting the discussion to only part of the ToT vs PTD discussion: e.g. the role of farmers' vs scientific knowledge.
- Ask the participants to form two groups, each of which will "defend" one of the two approaches in the debate, and ask two participants to volunteer to moderate the debate.
- Give the two groups 15–20 minutes to develop their "case", i.e. to find arguments to defend their side of the debate.
- At the same time, the two moderators prepare themselves for the debate; specifically, how to manage the conflict built into the game so that it contributes to increasing the participants' understanding. Suggestions such as the following may help the moderators in their preparation:
  - try to identify related or opposing concepts in the discussion;
  - on the board, visualise these and other main points raised;
  - identify and define the possible conflicts;
  - try to avoid people taking sides in a destructive manner; but
  - also avoid trying to harmonise opposing views/forces;
  - the challenge is to tap the creative power between opposing views.
- Allow the debate to continue only for as long as it remains interesting – probably no more than 45 minutes.
- Moderate a plenary evaluation of the game, in two parts:
  - the main learning points on the content of the debate;
  - how does the tension between opposing views/forces help us, when managed well, to develop our understanding, and what does this mean for our work with farmers?

Variations:
- The debate could also take the form of a court case, with the judge playing the role of moderator. Numerous other variations in form are possible – use your own creativity!
- A controversial statement can be given for debate and participants asked to take sides in the debate on the basis of their own feeling towards this statement. This brings the discussion much closer to the heart of the participants. A possible statement: "Without the decisive influence of farmers in developing agricultural technologies, sustainable agriculture will never be reached".
- If the two groups are too large to permit effective preparation, the preparation can be done in smaller subgroups which still enter the debate as one "party". Alternatively, the debate can be held in two parallel groups, with only the final evaluation in the plenary session; two more participants then have an opportunity to practise moderating such a discussion.

Source: based on idea in
Salas, Scheuermeier & Gottschalk (1989)

LEARNING ACTIVITY 2
CASE STUDIES IN TECHNOLOGY DEVELOPMENT (TD)

about 3–4 hrs

Objectives:
- To enhance participants' ability to distinguish between different approaches to TD and assess their major potentials and limitations.

- To increase participants' commitment to farmers' participation during all stages of TD.

Setting/approach:
- In small teams, in a workshop setting, participants analyse different examples of TD and draw main lessons from them. If field visits are included, the activity can take up to one day.

Materials:
- Pen and paper, large newsprint sheets and markers; handout with assignment for case analysis (see p. 47); also handouts with case descriptions (if cases cannot be generated and prepared with the participants).

Procedure:
- Explain the objectives of this activity in terms of drawing lessons from detailed analysis of TD examples.

- Ask the participants to form teams of 3–5 persons to study the case according to the following procedure:
  - the participant acting as resource person briefly introduces the case: background, process, outcome;
  - other team members ask questions for clarification, in order to gain a complete picture;
  - the case is analysed following the guidelines in the handout.

- In selecting the cases, consider the following:
  - if the case is based on one of the participants' experiences, s/he will be the resource person in the small group; this permits discussion of a wide range of aspects;
  - a case focused on only one technology – such as animal traction, row planting or a food-processing technique – is easier to analyse;
  - most lessons can be drawn if the case covers both the development and subsequent "introduction" of this technology, preferably over several years;
  - it should be possible to recognise elements of at least one of the three TD modes in the case.

- After about 1½ hours, ask the teams to hang their sheets with results on the wall.

- Each case is then presented and discussed in turn. The team explains the main points written on its two sheets. The plenary discussion is focused on understanding the main weak and strong points of the approach taken in that case (30–60 minutes per case).

- Ask the participants to identify individually three main lessons drawn from analysing the cases, and collect these lessons on the main board.

Variations:
- The teams may also be given more specific questions directly related to the issues raised in the discussion section above, for example, to what extent did farmers participate? Why was farmers' participation encouraged? Which elements of the three TD approaches do you recognise?

- If suitable cases can be found near the location of the workshop, the case studies can be expanded to include visits to the sites and interviews with both farmers and officials.

HANDOUT 1·1

ANALYSIS OF TECHNOLOGY DEVELOPMENT CASES
Assignment for Groupwork

1. Analyse the technology development case presented

    - by distinguishing the following four steps:
        1. Getting started and defining the problem
        2. The actual process of technology development
        3. Communicating the technology to farmers
        4. Follow-up and further developments

    - by answering the following questions for each step:
        a. What actually happened, in brief?
        b. What was the role of farmers in this step?
        c. What was the role of outsiders (NGO, extension service, etc)?
        d. What were the main problems encountered?

2. Formulate conclusions on the basis of the above analysis:

    - To what extent was the technology development successful?
    - What are the strong and weak points of the approach? Why?

3. For plenary discussion, prepare a presentation consisting of:

    - One newsprint sheet summarising the results of the analysis in Point 1.
    - One newsprint sheet with the main conclusions drawn in Point 2.

LEARNING ACTIVITY 3
PTD vs ToT IN WEST AFRICA

2 hrs

Objectives:
- To enhance participants' ability to distinguish between different approaches in technology development and to assess their major potentials and limitations.
- To increase awareness of their own (organisation's) strategy towards technology development as compared to PTD and ToT.

Setting/approach:
- Showing of a short slide series, followed by analysis in small groups.

Materials:
- The World Neighbors slide series on Community-Based Experimentation and Extension (CBEE)* with script; materials for presenting groupwork results in the plenary session.

Procedure:
- The slide series is shown in a plenary session; the script may be read by two participants taking the roles of the two extension workers.
- The first 9 slides are shown, presenting the experiences of one extension worker; in pairs, the participants then reflect on the following questions:
  - Why is this extension worker frustrated? What is happening in his work?
  - Why do you think the farmers are behaving like this?
- After discussing the results of this buzz session in the plenum, the participants reflect jointly on the questions: "Do we recognise this experience?" (or "Do we experience similar things in our own work?") and "What needs to be done?".
- The rest of the slide series is presented; in small groups of 3–4 persons, the participants then reflect on the following questions:
  - What are the main differences between the approach of the first extension worker and that of the second?
  - Why does the second approach have more effect?
  - How applicable is this approach in your own work situation?
- In the plenary, the answers are presented, compared and discussed. Issues that may arise are: the need for participation in order to have real impact, participation in various stages of the process, and differences between a participatory approach and ToT.

Variation:
- Prior to watching the slides, participants may be asked to gain background information by reading the article on the CBEE approach in *ILEIA Newsletter* 4 (3): 11–14.

LEARNING ACTIVITY 4
THE IPRA METHOD: A VIDEO

2 hrs

Objective:
- To increase participants' awareness of the potential of a high degree of farmers' participation, especially in agricultural research.

Setting/approach:
- Showing of a video film, followed by analysis of the case in small groups.

Materials:
- Video "The IPRA Method";* materials for presenting groupwork results in the plenary session.

Procedure:
- The video "The IPRA Method" is shown in a plenary session. It describes participatory research in the Andes in which staff of the international agricultural research centre CIAT were involved.
- Participants are asked to answer the questions in the study guide that accompanies the video. Many variations are possible in analysing the video; some work individually or in small groups.
- In the final plenary session, results of the individual or group work are presented, compared and analysed. The central issue in the discussion within the context of this unit would be the comparison between the IPRA method, PTD, ToT and the participants' own approach.

Variation:
- Another video, "Participatory Research with Women Farmers",* may be viewed and studied in a way similar to the above. This describes the approach of another international agricultural research centre, ICRISAT, in the drier parts of southern India.

* *For details, see "Audiovisuals and Simulation Games" in the Appendix to this guide, pp. 222–3.*

UNIT 1·D / PTD: FRAMEWORK AND KEY FEATURES

OVERVIEW OF THIS UNIT

EXPECTED RESULTS

The PTD approach is depicted as a series of activities grouped as stages within a process. However, this does not mean that these specific groupings and this phasing constitute "The Recipe". They represent, rather, a generalised summary of what farmers and development agents working together in various parts of the world have done, before the term "PTD" was created to describe it. The systematisation of actual experiences helps in understanding the importance of the different groups of activities and their contribution to the participatory approach as a whole. It also helps in understanding the key features of the approach as a basis for adapting it to one's own situation.

After completing the learning activities in this unit, the participants are expected:

- to understand the PTD approach as a nonlinear, iterative process consisting of various possible combinations of activities, rather than a series of predetermined steps;
- to be aware of the different groups of activities within a PTD approach and of their contributions to the entire PTD process;
- to understand the key features of PTD and be able to define their own position towards them.

MAIN CONCEPTS

- **PTD framework:** A presentation of a series (of groups) of activities undertaken to realise a PTD process.
- **Portfolio of options:** Several ideas for improving farming are developed or offered for farmers to select from and "play with" in various combinations (contrary to transfer of a technology "package").
- **Local experimentation capacity:** The knowledge and skills of farmers and communities to develop new practices and insights through series of experiments.
- **Roles of fieldworker:** A combination of facilitator, networker, educator and co-researcher.

TRAINING METHODOLOGY

As this unit is meant to help participants develop concepts, it involves mainly group discussions based on case studies of PTD programmes, either presented with audiovisual aids or visited in the field. Most learning activities incorporate some form of critical reflection on participants' own views of the issues at hand. The discussion section of the unit can serve as an input for such reflection; alternatively, opinions and views collected from the participants can be used. Many elements of PTD discussed here are treated in more detail in other units, as indicated by references in the text.

LEARNING ACTIVITIES

1. Phases of a PTD process
2. Towards a PTD framework: case studies
3. Key features of PTD: digging deeper
4. Talking positively about PTD
5. Obstacles to participation

DISCUSSION

THE PTD FRAMEWORK

PTD encompasses activities and methods to encourage participation of men and women farmers in developing agricultural technologies. These activities and methods are clustered in six themes (Table 1.4, pp. 50–51) which, together, form the PTD framework. These themes will be dealt with in more detail in other units as indicated. The main focus here is on the logic of the framework, i.e. of the process as a whole, and the coherence between the different clusters.

Clearly, "technology" in PTD should not be understood only as crop varieties, agricultural tools, husbandry practices or farm plans. Technology also includes mental constructs; it embodies certain cultural codes and forms of management and cooperation. Outsiders' support to the development of local-level organisations likewise follows a PTD approach of joint analysis, experimentation, learning and consolidation.

In practice, one rarely finds a programme following all six themes one after the other in a linear sequence. Some programmes focus on only one theme; others move back and forth between themes, e.g. between analysis and experimentation. Through such a process, the understanding of all participants is deepened, and new issues and activities arise.

TABLE 1·4
THE PTD FRAMEWORK

Clusters	Rationale	Elements	Expected Outcomes
Getting started (see Unit 2-A, pp. 89–101)	Taking a participatory approach does not mean starting PTD initiatives unprepared. Several important issues need to be addressed before intensive interaction with farmers can begin.	■ receiving a request to start collaboration, or selecting communities with which collaboration will be sought; ■ gathering and analysing existing secondary data; ■ making an inventory of existing organisations; ■ clarifying one's own agenda and possibilities for follow-up after situation analysis; ■ building a relationship with the local people and coming to a basic agreement on the form of future collaboration.	■ a clear perspective and protocols for collaboration; ■ a preliminary understanding of the socio-cultural and agro-ecological situation of the community or communities; ■ a core network of individuals and organisations that could play an important role in future PTD work.
Understanding problems and opportunities (see Units 2-B & 2-C, pp. 102–12 & 113–28)	The strongest driving force of a participatory programme is the farmers' realisation that it really addresses their particular concerns. A joint understanding of these concerns must be developed. At the same time, ideas for innovation already present among the farmers may provide good opportunities for commencing PTD.	■ sharing impressions of trends and problems in local farming; ■ supporting farmers in identifying and analysing their problems and the cause–effect relationships involved; ■ clarifying whose problems have been identified; ■ discussing the context of the problems (e.g. wider agro-ecological system, socio-political changes) and analysing driving/restraining forces; ■ making an inventory of opportunities and potential resources, including human resources and good ideas. The PRA (Participatory Rural Appraisal) toolbox is an important source of methods and techniques for these activities.	■ shared insight into local agricultural potentials and constraints; ■ improved skills of farmers to diagnose and analyse problems; ■ increased self-confidence and a better organisational basis for systematic experimentation by farmers.
Looking for things to try (see Unit 2-D, pp. 129–40)	Research and extension agencies are not the sole source of innovations to solve the problems or tap the opportunities identified. Also farmers and artisans in the community or elsewhere can provide interesting ideas to follow up. The various ideas are screened systematically by the farmers and PTD facilitators, and a joint agenda for experimentation is developed.	■ gathering information for detailed analysis of the identified concerns and priority problems; ■ identifying promising solutions from local experience, farmer experts and sources outside the community; ■ making a critical review of the options by establishing criteria for selecting initial activities and assessing advantages and disadvantages; ■ clarifying expected effects of the options on different subgroups within the community and the area; ■ developing an understanding of the need to experiment with the options selected; ■ agreeing on what exactly is to be found out by doing the experiment (formulating the hypothesis to be tested).	■ overview of possibly relevant technologies; ■ agreement on the most interesting option(s) to be tried out; ■ improved linkages between farmers and sources of innovations.

TABLE 1·4 (continued)

Clusters	Rationale	Elements	Expected Outcomes
Experimentation (see Part 3, pp. 141–88)	The focus is on experiments that farmers can manage and evaluate themselves and that give results on which the farmers can base sound decisions. Through involvement in these activities, farmers improve their capacity to adapt their agricultural practices. This is achieved through skill development, group building, and strengthening exchange and supportive linkages with other communities and organisations.	■ reviewing farmers' existing experimental practices; ■ designing selected experiments; ■ defining evaluation criteria and choosing monitoring and evaluation tools; ■ training farmer-experimenters; ■ establishing and managing the experiments; ■ monitoring by the farmer-experimenters supported by PTD facilitators; ■ evaluating results, both during the course and at the end of the experiments, to decide if the option is suitable locally, to develop possible technical guidelines for applying it and/or to identify any need for further experiments; ■ reviewing the experience of collaboration and experimentation with a view to improving the PTD process.	■ insight into the functioning and value of innovations, gained through experiments planned, implemented and assessed by farmers; ■ development of technology adaptations that are relevant locally; ■ improved capacity and skills of farmers in experimentation; ■ increased understanding of PTD processes.
Sharing the results: farmer-based extension (see Unit 4·A, pp. 192–206)	Many of the above activities involve farmers learning from other farmers – while discussing problems and opportunities, seeking good ideas and analysing results of experiments. PTD also encourages wider sharing of results among other farmers, using the networks developed during earlier PTD activities. Not only are the locally-developed technologies disseminated, but attention is also given to sharing the methodological aspects of learning through experiences of farmer organisation and experimentation.	■ studying the existing patterns and channels of farmer-to-farmer exchange and learning; ■ strengthening farmer-to-farmer exchange: visits, farmer-to-farmer training through learning-by-doing; developing manuals and audiovisuals by and for farmers; ■ training farmers as grassroots extensionists/promoters.	■ enhanced farmer-to-farmer diffusion of ideas and technologies; ■ building up an inter-community PTD network; ■ involvement of an increasing number of communities in systematic technology development; ■ establishment of a farmer-managed system of inter-community training and communication.
Sustaining the PTD process (see Unit 4·B, pp. 207–13)	The ultimate aim is to leave communities with a capacity to implement an effective process of change. PTD programmes are therefore concerned with organisational development and the creation of favourable conditions for ongoing experimentation and development of sustainable agro-ecological systems. The role of outside PTD facilitators gradually changes. Their attention begins to shift to other communities and areas. They gradually phase out their support at one site, in order to be able to promote PTD on a wider scale.	■ stimulating group development and linking groups with farmers' organisations; ■ providing training in fields related to management; ■ strengthening linkages between (groups of) farmers and service organisations; ■ consolidating institutional and policy support to PTD processes; ■ documenting the process and methods of experimentation and diffusion; ■ supporting evaluation of the impacts of technologies and the PTD process on the livelihood system.	■ consolidated community networks or organisations for agricultural self-management; ■ a more supportive institutional environment; ■ documented and operationalised PTD approach and resource materials; ■ relevant services and input supply.

KEY FEATURES

The key features of the PTD approach described above are:

- It encompasses all elements of the *overall technology development process*. It goes beyond appraisal, situation analysis and setting an agenda for action to include experimentation, evaluation, sharing and consolidation.

- It provides a clear *link between farmer-led research and farmer-led extension*, thus integrating research and extension at the farmer level instead of linking these only at the level of formal institutions.

- It recognises and respects in all stages the importance of *men's and women's indigenous knowledge*. Such knowledge is not seen as static. It changes over time, also through deliberate efforts of farmers to try out new ideas. One of the major challenges in PTD is to build bridges between farmers' knowledge, which is holistic by nature, and specialised scientific knowledge (Salas *et al.*, 1989).

- It focuses on *farmer-led experimentation*, rather than demonstration and adoption of innovations. It is during experimentation that farmers' own knowledge and experience are brought together with outsiders' insights, and are compared and analysed to arrive at a locally-appropriate synthesis.

- It aims at *enhancing farmers' capacity* to develop farming systems that are sustainable over time and which conserve and improve local resources. It increases farmers' resilience to changes in their circumstances.

A participatory approach does not guarantee a move towards more sustainable forms of agriculture, nor does it automatically alleviate poverty. Existing power relations – of men over women, of the rich over the poor, of the old over the young – will affect the participation process and prevent equal benefit to all. Conventional agricultural research and extension used to be biased towards male, wealthier and better educated farmers. If PTD is to avoid such biases, certain methods and tools must be applied. These are discussed at appropriate points in Parts 2, 3 and 4. Units 1.E & 1.F – "Respecting Rural Life" and "Gender Sensitivity" – generate the basic understanding needed to make good use of these methods and tools.

ROLES OF FIELDSTAFF IN PTD

The process of PTD assigns roles to fieldstaff which differ markedly from those played in conventional agricultural research and development. The new roles of fieldworkers are initially as:

■ Facilitators

- in analysing the present farming situation and resource base;
- in making an inventory of local knowledge and ideas which could provide solutions for identified problems, and other local resources relevant for developing a sustainable farming system;
- in making farmers' criteria explicit and selecting options to try;
- in making farmers' informal experimental methods explicit and in systematic planning, monitoring and evaluation of new experiments;
- of farmers' self-organisation and self-management.

■ Networkers

- by encouraging exchange among farmer-experimenters in the local area and beyond;
- by helping to develop linkages with local farmer organisations and relevant support organisations;
- by linking farmers with relevant sources of information;
- by feeding back information about farmers' experiments to formal researchers;
- by stimulating greater participation of farmers in programming and assessing formal research.

■ Educators and trainers

- by enhancing farmers' diagnostic capacities;
- by revitalising indigenous knowledge, cultural identity and self-esteem in farming communities;
- by helping farmers increase their understanding of the principal processes at work in their agro-ecological system;
- by facilitating the development of relevant organisational and communicating skills (such as problem-solving, leadership development and functional literacy) of farmer-trainers and PTD group leaders.

- **Co-researchers**
  - by contributing ideas, potential solutions and information from formal research, whenever useful in the PTD process;
  - by replicating farmers' experiments under more controlled conditions;
  - by making additional measurements and observations to support analysis of farmers' experimentation;
  - by documentating the entire process and the final evaluation of socio-cultural and agro-ecological impacts.

As the PTD process develops, fieldworkers gradually assume the role of:

- **External advisors**
  - by participating in the evaluation and planning of group activities at the farmers' request;
  - by supporting leaders of farmer-experimenter groups;
  - by assisting in networking and policy lobbying at inter-community and higher levels.

OBSTACLES TO PARTICIPATION

In promoting farmer participation in technology development, many obstacles have to be overcome. Those most commonly encountered are the following:

- Local government agencies and bureaucratic forces, despite their rhetoric of support, have reasons to fear farmer participation and may seek to divert the threat. They may appear to accept the participatory programmes, but then take them over and give them a completely different meaning.

- Some professional agronomists and development workers find it hard to accept that rural people have something to contribute to technology development. Through many years of formal education, they have been led to believe that scientific knowledge is superior to local knowledge. This prejudice is very difficult to overcome.

- Many organisations, both governmental and non-governmental, lack the flexibility and the internal openness to follow a participatory approach. Where bureaucratic or charismatic leaders dominate and dictate the day-to-day work of their staff, there is little hope that the latter can develop strong participatory interaction with their "target group".

- A large part of the rural population – women – face special obstacles: heavy labour inputs prevent them from taking part in meetings; cultural restrictions prevail against appearing or speaking at open meetings; their expertise and independent interests are easily neglected in community action. Deviation from the norm, which is implied in experimentation with new ideas, sometimes raises very strong opposition.

- In most countries, there are disadvantaged minorities which are distinguished by race, religion or ethnic group. The participation of these minorities in development activities may be strongly resisted by the dominant groups.

- The poverty of certain categories of the rural population and their previous bad experiences with (non-) supporting agencies may have robbed them of any hope for improvement, depleted their self-confidence and increased their distrust of outsiders. This results in a "culture of silence".

LEARNING ACTIVITY 1
PHASES OF A PTD PROCESS

2–3 hrs

Objectives:
- To enhance participants' understanding of the PTD approach as a nonlinear, iterative process, consisting of various possible combinations of activities rather than a series of predetermined steps.
- To make participants aware of the different groups of activities within a PTD approach and their contribution to the PTD process as a whole.

Setting/approach:
- Essentially a small-group activity followed by a plenary discussion to draw joint conclusions.

Materials:
- For presenting results of the small-group discussions, a board (or newsprint), small cards of different colours, and markers.

Procedure:
- Ask the participants to form small groups of 3–5 and give them the following instructions:
  - List important phases in a PTD process according to your own opinion or experience: what needs to be done first? what next? and so on;
  - For each phase identified, describe briefly:
    a. why it is done (the objectives)
    b. how it is done (the methods)
    c. the roles of farmers vs "outsiders" (PTD facilitators);
  - Prepare a visual report using the small cards – those with one colour for the phases, with another colour for the objectives, and so on.
- Invite the groups, in turn, to present the results of their discussion, allowing only questions of clarification after each presentation.
- Facilitate a plenary discussion of all the groups' results by posing general questions such as "What important differences are there between the results and why do such differences occur?" and more specific questions such as "Why did one group not include … whereas another group did?" It should be stressed that the phases of PTD can be presented in several different ways, and that the process can take a zigzag path with many repetitions. The discussion may (but does not have to) lead to agreement among the participants on a common framework to be used during the rest of the training.
- Finally, show participants the PTD framework (Table 1.4, pp. 50–51) and ask if it reveals any issues overlooked in the discussion up to now.

Variations:
- Instead of being asked for opinions about what form a participatory process should take, participants could be asked to describe how technology development is currently taking place within the context of their work. Comparing their descriptions against each other (and possibly against the PTD framework in Table 1.4) may confront participants with shortcomings in their own approaches.
- Learning Activity 2 uses one or more concrete examples of PTD programmes to study the above issues. If such cases are available, they may be especially useful in helping groups which have difficulty in working on a purely conceptual level.

LEARNING ACTIVITY 2
TOWARDS A PTD FRAMEWORK: CASE STUDIES

2–3 hrs

Objectives:
- To enhance participants' understanding of the PTD approach as a nonlinear, iterative process, consisting of various possible combinations of activities rather than a series of predetermined steps.
- To make participants aware of the different groups of activities within a PTD approach and their contribution to the PTD process as a whole.

Setting/approach:
- Essentially a small-group activity in which concrete PTD cases are studied in detail.

Materials:
- Handout with case example(s); for presenting results of group discussions, either small cards or newsprint with markers. Cases developed from local experience are the most effective, but ones included in this book can also be used; for example, the CBEE case studied in Learning Activity 3, Unit 1.C (p. 48), or the Sri Lanka case presented in Box 3.5 (p. 163). The *ILEIA Newsletter* is another potential source of case descriptions, e.g. Vol. 4 (3), Vol. 10 (2) or Vol. 11 (10). Studying more than one case shows that there is no single way of doing PTD and challenges participants to define their own position.

Procedure:
- Distribute the case(s) or watch the slide series.
- Ask the participants to form small groups of 3–5 persons to analyse the case(s) according to the following instructions:
  - List the most important steps in the approach: what was done first? what next? and so on;
  - For each identified step, describe briefly:
    a. why it is done (the objectives)
    b. how it is done (the methods)
    c. the roles of farmers as opposed to "outsiders" (PTD facilitators);
  - Prepare a visual report using the small cards: those with one colour for the main steps, with another colour for the objectives, and so on.
- Invite the groups, in turn, to present the results of their discussion, allowing only questions of clarification after each presentation.
- Facilitate a plenary discussion of all the groups' results by posing general questions such as "What important differences are there between the results and why do such differences occur?" and more specific questions such as "Why did one group not include … whereas another group did?" It should be stressed that the phases of PTD can be presented in several different ways, and that the process can take a zigzag path with many repetitions. The discussion may (but does not have to) lead to agreement among the participants on a common framework to be used during the rest of the training.
- Finally, show participants the PTD framework (Table 1.4, pp. 50–51) and ask if it reveals any issues overlooked in the discussion up to now.

Variation:
- Learning Activity 1 (p. 54) addresses the same issues but without the use of specific cases, i.e. at a more conceptual level.

LEARNING ACTIVITY 3
KEY FEATURES OF PTD: DIGGING DEEPER

3–4 hrs

Objectives:
- To increase participants' critical understanding of the key features of a PTD approach.

Setting/approach:
- A brainstorming session to collect participants' views on key elements is followed by an in-depth analysis of selected themes in small teams. Participants are expected to have experience with farmer participation and to be familiar with conceptual analysis.

Materials:
- Small cards and a board for the brainstorming; newsprint and markers for reports of the small-group discussions.

Procedure:
- Start the brainstorming by asking participants to think of two basic features of a farmer participatory approach to technology development, to write the essence of each issue on a card, and to hang the cards on the board.

- In a plenary discussion, group similar cards, e.g. begin with one card, ask if the second card fits to the first or is a separate issue, the third one fits to the second or the first or is a separate issue, and so on. For each group of cards, agree on a name that describes the common theme.

- The participants then form small groups to elaborate on the identified themes, selecting their subgroups according to their own interests. Each subgroup discusses various aspects and issues of their theme and records the main points of their discussion on the newsprint (45 minutes).

- Ask the subgroups to hang the newsprint sheets with their results on the wall for all to see; encourage the participants to walk around and read the results, and to reflect on them, perhaps discussing them informally with others reading the same sheet.

Variation:
- The results of the subgroups can be used in a final plenary session to stimulate deeper-going exchange of views and experiences. However, the previous steps may have demanded so much concentration that participants have little energy left for a plenary discussion.

Source: developed at the workshop on training for PTD (1990), Leusden

LEARNING ACTIVITY 4
TALKING POSITIVELY ABOUT PTD

45–60 min

Objectives:
- To sense the importance of having an exploratory rather than an analytical mind-set and of using positive, creative language when talking about PTD.
- To discover the most appropriate phrases referring to PTD in the local language.
- To encourage reflection by participants about how they really understand PTD.

Setting/approach:
- Work in small groups in a workshop setting.

Materials:
- Overhead with formulations relevant for introducing the idea of PTD in rural communities.

Procedure:
- Start the session by presenting handwritten overheads suggesting formulations (in English) which fieldworkers could use when introducing PTD activities in a local community (see examples in Box 1.3). These statements may need a brief plenary discussion.
- Ask the participants to translate key expressions such as "improving the situation", "skills and knowledge", "things that work" and "working together" into the local language of the area they work in. This can be done either individually or in groups of 2–3 persons. The translations should be written down for presentation in the plenum.
- The plenary discussion of the results should be aimed at gaining a joint understanding of the key expressions. What causes the differences in the choice of local terms? What do we really mean with PTD? This can be made explicit by asking participants to make a literal translation back into English of the local expressions. This may bring hidden connotations to light. A confronting question can be posed, such as: Who in the village would use these local words? Would a poor farmer use them or only the village chief?

Variation:
- The above statements are selected to encourage a change in participants' perspectives – from a problem-analysis reflex ("What is your problem and what are your needs?") to an exploratory approach ("How could this situation be improved?"). However, other statements coming out of the ongoing training process could also be selected for this activity.

BOX 1-3 TALKING POSITIVELY ABOUT PTD

SAY:
- "We want to discover opportunities for improving the situation."
- "We must understand the situation here, and nobody knows it better than you."
- "What could be done? How can we join forces to discover what can be done?"
- Repeatedly explain: "We want to combine our skills and knowledge with yours. Hopefully we can then jointly find new useful things that work. We want to do this, because we want our work to be useful to you. Otherwise, there is no reason for our work."
- "What is the situation here? What can be done about it? How can we join forces to do something about it?"

AVOID SAYING:
- "We have come to find solutions to your problems."
- "You must tell us what problems you have."
- "How can we help you?"
- Avoid talking of material inputs and money. When asked, explain that such things might be needed, but we are interested more in working together. If they are only interested in getting materials and money from us, then we are not interested in doing PTD with them.
- "What do you need?"

Source: Scheuermeier & Sen (1994)

LEARNING ACTIVITY 5
OBSTACLES TO PARTICIPATION

2–4 hrs

Objectives:
- To enhance participants' understanding of possible constraints to promoting farmers' participation in technology development.

- To promote participants' commitment to address some of these constraints which are within their reach to influence.

Setting/approach:
- This is a form of problem census, a method of analysing problems together with farmers as discussed in Unit 2.C (pp. 113–28). Work in small groups is alternated with reflection in the plenary session. Participants may be asked to moderate parts of this activity, to give them an opportunity to practise this method.

Materials:
- To visualise the results of groupwork, preferably small cards, markers and a board to which the cards can be attached.

Procedure:
- Outline the general procedure of this activity and explain, if necessary, how to use the cards.

- Ask the participants to form small groups of 3–5 persons to:
  - brainstorm on experienced constraints in promoting farmers' participation and/or foreseen ones;
  - analyse each and write the essence of the constraint on a card.

- Facilitate a plenary discussion of the results:
  - each group explains its cards;
  - common themes are identified by clustering cards with similar problems; find a label for each cluster that expresses the common theme, the central constraint.

- Ask the participants to go back into their small group to:
  - find the one or two most important constraints;
  - brainstorm on possible ways to overcome these.

- In the final plenum the developed suggestions are exchanged. Possible implications for the rest of the training may emerge.

UNIT 1·E / RESPECTING RURAL LIFE

OVERVIEW OF THIS UNIT

EXPECTED RESULTS

Understanding and respecting rural life, the way farmers manage their farm and household in often difficult situations, the complexity of the farming systems they have developed – without these, a PTD practitioner will not be able to collaborate with farmers in the way indicated in the previous units.

After completing the learning activities in this unit, participants are expected:

- to understand the complexity of farmers' livelihood systems and the close links between the different enterprises;
- to respect the value of farmers' knowledge in managing their natural resources;
- to be aware of the various objectives farmers are trying to reach, as compared with the few that usually receive attention from outsiders;
- to be aware of different forms of community organisational structure and the power relationships involved, including their own role in these;
- to have developed an attitude of respect for how farm families manage their lives.

MAIN CONCEPTS

- **Livelihood system:** The combination of many different, yet often intimately integrated, activities or enterprises that form the basis of the survival and well-being of farm families.
- **Indigenous knowledge:** Knowledge generated locally, or derived from elsewhere but transformed by local people and incorporated into their way of life.
- **Farmers' objectives:** The reasons for the activities of farm families, going beyond productivity, to include perhaps continuity, stability, equitability or other (often unquantifiable) values for which they strive.
- **Community organisational structure:** The various informal or formal organisations and collaborative patterns existing within the community.
- **Power relationships:** Unequal distribution of influence on decision-making within the community, which can strongly influence possibilities for PTD.

TRAINING METHODOLOGY

This unit focuses mainly on attitudinal change towards greater respect for the lives of small-scale farmers. A very effective way to achieve this is to become immersed in or confronted with farmers' reality, for at least a few days but maybe even a month or more. Training in PRA methods and tools (Unit 2.C, pp. 113–28) provides good opportunities for this. Simulation games have also been designed to allow participants to experience farmers' reality within a workshop setting, e.g. the game "Africulture". Also, articles such as "The small farmer is a professional" (Chambers, 1980) can be read and discussed to support the other activities in this unit. As respect is closely linked with understanding, several "classroom" activities can be combined with real-life confrontation to enhance understanding. Unit 1.B (pp. 30–40) on principles and methods of LEISA also suggests learning activities aimed at gaining a better understanding of the complexity of farming systems.

LEARNING ACTIVITIES

1. Immersion in village life
2. Storytelling about local customs and beliefs
3. Understanding others
4. Analysis of village society
5. Experiencing power play: the game "Objective"

The simulation game "Africulture", available from ETC, comes with an extensive trainers' manual (for details, see p.222: ETC).

DISCUSSION

THE COMPLEXITY OF FARMERS' LIVELIHOOD SYSTEMS

When an outsider observes rural communities and the activities taking place there, there is a temptation to look down on farmers' way of life and their efforts to survive, which superficially appear not to be very successful. Many professionals working in agricultural research and extension have, in fact, been trained for years to regard farmers in this way. It is only when one takes a real interest and makes an effort to understand their day-to-day struggle that one begins to respect how farmers manage to make the best of often extremely difficult situations, and to recognise that they have developed effective, integrated strategies to reach their aims in life. Working in a participatory way with rural communities requires an understanding of, and respect for, these achievements.

As a first step in developing this understanding, fieldworkers need to see and comprehend how local livelihood systems combine many different, yet often integrated, enterprises: not only farming activities (crop, tree and animal husbandry) but also home-based production (food processing, handicrafts, carpentry, etc) and sometimes off-farm employment. Reasons for this diversification may include:

There is a need to understand the community's activities and achievements

- spreading risks and thus ensuring at least a minimum of food and income if disaster strikes one or more of the enterprises;
- mutual support between the activities: output from one serves as input for another, thus increasing total efficiency of resource use;
- better use of time, by avoiding strong peaks in labour and filling in otherwise slack periods.

INDIGENOUS KNOWLEDGE

In managing these multiple activities, farmers rely on their intimate knowledge of the local situation. This is often referred to as "Indigenous Knowledge" (IK): the ideas, experiences, practices and information that either have been generated locally or are generated elsewhere but have been transformed by local people and incorporated into their way of life. IK is deeply embedded in the local cultural, social and economic context. It is not abstract like scientific knowledge; it is concrete and relies strongly on intuition, historical experiences and directly perceivable evidence.

Realising that IK differs from the formal scientific knowledge system is the starting point of PTD. The differences between the two knowledge systems are due to differences in concepts, tools, skills, environments and priorities. For example, in Andean culture, IK is linked with the idea of reciprocity with Mother Earth, whereas agricultural scientific knowledge considers land as an economic factor of production and seeks to maximise its use.

IK is not static, nor does it depend only on local ideas. The community also absorbs, transforms and internalises ideas from outside, so that they become a part of their IK. Respecting IK is the key to partnership, as it gives dignity to the local farmers and puts them on an equal footing with the outsiders involved in the process of technology development. The fact that IK is far from uniformly distributed within or across communities (Scoones & Thompson, 1994) has important implications for PTD, as will be elaborated below.

FARMERS' CRITERIA

Researchers and extensionists often see agricultural development as a quest to maximise production, but this is only one of a number of objectives pursued by farmers – even those in well-endowed areas with good access to markets. For this reason, the priorities of farmers and outsiders, and the way they assess new technologies, can differ considerably. Even if only those objectives that are directly related to farming are considered, they are likely to be many and varied. Conway (1987) grouped the various objectives of farmers as follows:

- **productivity:** the output of products per unit of land, labour, capital or unit of input, and per unit of time. Output in this case is not only evaluated in terms of market value (total yield times market price) but also on the basis of criteria such as storability, taste, nutritional value, cooking quality and resistance to pests;

- **continuity:** the ability to maintain productivity and thus survival in the face of a major disturbing force (e.g. erosion, declining prices, flood). Continuity is often also linked to conservation of natural resources for the farmers' children;

- **stability:** the ability to maintain productivity in the face of small disturbing forces arising from the normal fluctuations and cycles in the environment. To manage to do this, most farm families have strategies to minimise and balance the risks caused by such disturbances, by diversifying in space and time;

- **equitability:** the evenness of distribution of the productivity among the human beneficiaries, in the household, the community and possibly also beyond. This may also be related to the evenness in satisfying social and cultural needs such as status, honour, family ties and ethnic identity.

Farmers have to weigh these objectives in their daily decision-making. For example, a farmer may decide to participate in a three-day ceremony at his cousin's village rather than transplanting already overmature rice seedlings, if he feels that maintaining a good relationship with his cousin is, in the long run, more important than obtaining a satisfying rice yield in this particular year.

COMMUNITY ORGANISATIONAL STRUCTURE

To be able to understand farmers' decision-making, outsiders also need to be aware of the farmers' position and role in the community organisational structures. A farmer will generally belong to various units or groups, e.g.:

- **household** – the people who regularly live together within a dwelling: farm households usually consist of more than just a nuclear family (father, mother and children) and include grandparents, cousins and other relatives and dependants. Many day-to-day decisions are taken within the household, with different family members having different influence, depending on their age, sex, etc;

- **household network:** many farmers work together and exchange goods and services "freely" within a more or less well-defined network of several households, particularly for larger agricultural operations, house-building, etc. When making decisions related to these activities, members of the network tend to consult each other;

- **clan or larger kinship structure:** these larger family structures generally decide on and manage cultural and religious activities, and traditionally play a role in maintaining and defending local law and order;

- **local development organisations:** groups of individuals with common interests for improving the livelihoods of their families and communities, developed either indigenously or with support from outsiders. There may be several organisations active in one village (e.g. cooperative society, women's organisation, traditional savings-and-credit group), and some may imply automatic membership for all farmers;

- **regional or national organisations:** these include government organisations, religious organisations, political parties, etc.

CONFLICTING INTERESTS

Within and between the various groups, conflicts are certain to occur: who decides on what? who benefits? This will depend greatly on the social and economic status of the individuals or groups: nobility vs (former) slaves, large landowners vs farm labourers, mobile pastoralists vs settled farmers. Also within families, conflicts will occur. The limited influence of women in many development activities has been widely documented and will be discussed specifically in the next unit. But there are also differences in influence between the generations in families and clans.

PTD practitioners need to be sensitive to the power issues and their implications in such complex socio-cultural settings, and to recognise spheres of influence and how particular groups or individuals are being favoured or disadvantaged. As soon as PTD practitioners decide to work in a community, they cannot avoid becoming part of this "power play". Their efforts will be evaluated, at least in part, according to the extent to which they help the one or the other party.

Indeed, PTD practitioners with their support organisation represent a new organisational structure in the local setting, and one with relatively great influence. To handle this power issue in a responsible way is a major challenge for any PTD team. Becoming aware of it is the first important step. Units 2.C and 2.D (pp. 113–40) indicate how fieldworkers can learn about local situations from this perspective.

CULTURAL IDENTITY

Respect for farmers' life is not limited to recognising their achievements in their struggle to survive and prosper in their own terms. It includes a respect for their culture, though outsiders may never fully understand it. This, on its own, is a major reversal: from often urban-based contempt for rural cultural expressions to deep respect for farmers' cultural integrity and also to re-evaluation of one's own cultural background.

The basic issue at stake is the ability of PTD practitioners to see village life, farmers' situations, their problems and the decisions they take from a villager's perspective without losing their outsiders' asset of sharpness in observing and analysing.

LEARNING ACTIVITY 1

IMMERSION IN VILLAGE LIFE

several days or weeks

Objective:
- To enhance participants' general understanding of and respect for farmers' life.

Setting/approach:
- Workshop, plus several days to several weeks spent in villages.

Materials:
- Arrangement with village authorities and villagers for food and accommodation, with some form of compensation, such as labour.

Procedure:
- Organise a preparatory workshop of 1–3 days, which may include several other learning activities in this unit. The final sessions are devoted directly to preparing participants for their stay in villages, including the drawing up of assignments to be completed during the stay.

- Arrange that the participants live in the villages for a period of, for example, two weeks, during which they complete the agreed assignments, such as:
  - live with a family and share the daily work;
  - involve yourself in as many agricultural operations as possible;
  - learn a particular agricultural skill from a farmer: select an interesting technique, ask the farmer to teach you how to do it, do it often, try to discover the advantages yourself, ask the farmer what s/he sees as its advantages, and prepare a report on it (to be shared later with the other participants in the training);
  - learn a particular household-related skill from a family member, following the same procedure as above;
  - while living in the village and sharing farmers' work, pay special attention to certain issues (but don't do a survey!): culture and customs related to agriculture, community organisational patterns, power plays and their actors, etc.

- During a second workshop, the participants exchange reports on their experiences and the particular agricultural and household-related skills they have learned. Encourage the participants to draw conclusions from their confrontation with village life. Finally, ask them to formulate further learning needs, now that they have experienced village life.

Variation:
- This immersion activity can be integrated with learning activities on specific methods and tools for situation analysis as presented in Unit 2.C (pp. 113–28).

Source: based on the approach used by the Centre for Youth and Social Development, Orissa, India (Jagadananda, personal communication)

LEARNING ACTIVITY 2
STORYTELLING ABOUT LOCAL CUSTOMS AND BELIEFS

20–200 min

Objective:
- To enhance participants' general understanding of and respect for farmers' life.

Setting/approach:
- Open group discussion in which participants take turns in describing examples of local customs.

Materials:
- At the most, some materials to visualise main points of the examples.

Procedure:
- Ask the participants to share with the others examples, from either their work or their own village, of certain customs and beliefs in agriculture that do not make much sense to a scientifically trained outsider but, in fact, are very functional.

- Maybe start by telling a story from your own or someone else's experience, such as: On an island in eastern Indonesia, villagers used to wait with harvesting of their maize crop, even when they were hungry, until the local religious leader gave the signal to harvest by starting to do so himself. Apart from other reasons, this prevented the maize from being harvested at too early a stage, which would lead to considerable loss in nutritional value.

- After several examples like this have been collected, ask participants to share examples of customs and beliefs which, even if closely examined, do not appear to make sense or may even be harmful. Then challenge the other participants to brainstorm on possibilities for some rationale behind these stories.

- A final plenary discussion may be appropriate to evaluate together the participants' reactions when hearing the stories, in order to realise that most of us still often find it difficult to accept them.

Variation:
- Where possible, elderly villagers (men and/or women) may be invited to take part in such storytelling sessions and help the participants to explain the history of the customs and their role in village life.

Source: based on an idea from Werner & Bower (1982)

LEARNING ACTIVITY 3
UNDERSTANDING OTHERS

1 hr

Objective:
- To come to terms with the difficulties inherent in understanding and relating to people whose frame of reference and experience is far removed from one's own.

Setting/approach:
- Creative analysis of photographs, working in small groups, in any quiet meeting place that is sufficiently large.

Materials:
- A selection of photographs from around the world (at least twice as many pictures as participants) showing adults – working, playing, resting – in a variety of settings, both rural and urban. Most pictures should indicate a social and economic level similar to that of the group with whom the trainees will be working.

Procedure:
- Ask the group to examine the pictures and to try to imagine what a person in the scene is doing and feeling.

- Ask each participant to select a photograph and then to choose one person in the picture and try to enter into that person's life.

- Give the group 15 minutes to prepare a story about the person in the picture, to be related in the first person, as though they were that person.

- In pairs, the participants show their pictures and listen to each other's stories, and then reflect on the experience.

- In further reflection during the plenary session, central themes may be:

  - the difficulties in getting really "inside" the person they were portraying;

  - recognition that one's own projections, preconceived ideas and feelings make this difficult;

  - drawing a parallel to their own work, during which the same mechanisms occur, e.g. in collecting and evaluating information.

- In the plenary session, encourage participants to consider how to address these difficulties in practice; e.g. by maintaining an open attitude, respecting others, posing questions and making observations to learn about the lives and needs of others.

Source: Crone & St John Hunter (1980)

LEARNING ACTIVITY 4
ANALYSIS OF VILLAGE SOCIETY

3 hrs

Objective:
- To enhance participants' understanding of the main patterns of community organisation and awareness of their complexity.

Setting/approach:
- Intensive work in small groups, possibly combined with fieldwork to crosscheck participants' analysis.

Materials:
- Pen and paper, large newsprint sheet for presenting results of groupwork, handout (optional) on possible community organisational patterns (e.g. based on the discussion section of this unit, pp. 59–62).

Procedure:
- Ask the participants to form small groups of 3–4 persons and give them the following instructions:
  - describe existing organisations in the communities you work in: name, size, membership, type of activities; start with your own views and, if you wish, check the handout for additional points of analysis;
  - list the main factors that brought about this situation;
  - identify the main conflicts within the community: between whom and on what issues;
  - visualise your analysis on newsprint (total time up to 1½ hours).
- Ask all the small groups to present their results briefly to the plenum.
- Conclude with a plenary discussion, addressing questions such as:
  - What are the main points we learn when comparing the results of the groups (complexity, common gaps and sources of conflict, importance of informal organisations and collaboration)?
  - What role do or can we play as PTD practitioners within such complex organisational settings with conflicting interests between groups and individuals? Is there still scope for an approach focusing on the community as a whole?

Variations:
- The analysis can be much enriched if participants are given the opportunity to go to the villages concerned and find out to what extent the conclusions of their analysis are valid. Alternatively, certain groups or people from the community may be invited to the workshop to present their own analysis, as a crosscheck or input about other perceptions.
- Venn diagrams, such as the one shown on p.209, can be suggested as a tool to visualise the analysis; they provide an opportunity to show which organisations are more important or closer to daily life, and can be used in interaction with farmers.

Source: based on an example from the Centre for Youth and Social Development, Orissa, India (Jagadananda, personal communication)

LEARNING ACTIVITY 5
EXPERIENCING POWER PLAY: THE GAME "OBJECTIVE"

1–2 hrs

Objectives:
- To understand and evaluate the importance of farmer participation.
- To enhance awareness, through experience, of conflicting goals and interests and the impact these have on the powerless.
- To reflect on the role of the outsider in such conflictual situations.

Setting/approach:
- Uses the game "Objective" (can be played with groups of 12–20 persons.) Ample time is needed for reflecting on what the participants experienced during the game.

Materials:
- 3–4 blindfolds, pieces of rope to tie hands and feet, and a longer rope (about 10 m).

Procedure:
- Introduce the game "Objective"; 9 or 12 persons are asked to volunteer and they are divided into 3 or 4 groups of 3 persons each.
- In each group, one person is blindfolded, the hands of another person are tied, and the feet of a third person are tied. Groups stand in a circle, around which there is a rope connecting the hand-tied persons. Observers stand around the circle. Four objects are placed outside the circle. The assignment is for each group to reach the object nearest to them.
- In the first round of five minutes, no talking is allowed. The second round adds the possibility of talking. In the third, the observers may come in to help. There can be a fourth round, again without intervention, and a fifth with deliberate participative intervention.
- Subsequent reflection in the plenum can follow several steps:
  • How did each person feel: blind, hand-tied, foot-tied?
  • What did they actually do? What was the objective? What significance had each object?
  • What is the relation with reality? What could the blind, hand-tied and foot-tied persons symbolise?
  • How did the observers intervene? Evaluate their intervention in relation to people's participation. What differences were there between the rounds?
- Summarise jointly the main lessons from the experience: e.g. with reference to communities, different interests, different capacities, ways of intervention, the importance of free communication.

Source: developed in 1992 by FMD Consultants, Santpoorterstraat 17, NL-2023 DA Haarlem, The Netherlands

DRAWING BY R. ANSELME

UNIT 1·F / GENDER SENSITIVITY

OVERVIEW OF THIS UNIT

EXPECTED RESULTS

Men and women perform different roles in agriculture. They have different responsibilities and often different interests. Understanding these gender issues is of crucial importance in working with a participatory approach.

After completing the learning activities in this unit, participants are expected:

- to be aware of the different roles and responsibilities of men and women in agriculture-related activities and the need to take these into account in their work;
- to know which basic information on gender is critical in developing PTD strategies;
- to be able to integrate gender issues in their work.

MAIN CONCEPTS

- **Gender differences:** Differences between tasks and responsibilities of men and women, shaped by culture, norms and tradition in society.
- **Productive, reproductive and community work:** Work to ensure production of goods and services, care and maintenance of the household, and participation in community or political activities.
- **Access to and control over resources and benefits:** The ability to use natural resources and enjoy their benefits, and to influence decision-making regarding them.
- **Gender analysis:** An effort to understand and document the tasks and responsibilities of men and women as shaped by culture, norms and tradition.

TRAINING METHODOLOGY

Since participants always have some knowledge about differences between men and women in terms of their roles and the resources they use, the concept of gender can best be discussed by referring to their own ideas, experiences and knowledge. In the learning activities of this unit (pp. 71–4), several tools are suggested to encourage critical reflection on these ideas. Alternatively, the discussion section (pp. 67–70) can be distributed for joint analysis and discussion. The game "Africulture" (see Appendix, p. 222: ETC) incorporates many gender issues and may provide a good opportunity to start raising these issues in a new group.

LEARNING ACTIVITIES

1. Roles and behaviour of men and women: discussing stereotypes
2. Case study: "The Bean Experiment"
3. Strengthening an organisation to deal with gender issues

DISCUSSION

GENDER DIFFERENCES

Usually, the roles and responsibilities of men and women differ, according to norms and traditions in the society but also based on their social position. For example, in some cultures women and children collect water, while men collect fuelwood. This division of tasks and responsibilities is not based on biological differences (their sex) but rather shaped by the culture or society. This is called their gender. Roles and responsibilities are influenced by ideas and values about what is masculine and what is feminine and how men and women should behave. For example, in some cultures it is not considered proper that middle-class and high-class women work in the fields; they should remain in the family compound.

Some of these ideas about gender roles and expected behaviour are based on tradition; others are "imported" from other cultures. In other words, they are assigned and can change over time.

Women's tasks and responsibilities are not limited to household matters only, but cover three areas:

- **Productive work:** the production of goods and services for consumption and trade.
- **Reproductive work:** the care and maintenance of the household and its members, including bearing and caring for children, preparing food, collecting water and fuel, shopping, housekeeping and family healthcare.
- **Community work:** the collective organisation of social events and services such as ceremonies, community improvement activities, participation in groups, and local political activities.

Norms, values, ideas and social positions also influence the opportunity of different groups to make use of local resources: they do not have equal *access*

BOX 1-4 GENDER AND THE USE OF THE FOREST: THE CASE OF NEPAL

In a village in the Middle Hills of Nepal, the planting of permanent resources such as trees on private land is considered to be a man's job. The daily care of all livestock, by providing fodder and leaf litter for bedding, is the work of younger women and children. Collecting forest products, firewood, grass and leaf litter is thought of as women's work, while cutting firewood with an axe for seasonal use is considered to be men's work.

Decision-making and knowledge of decisions about the exchange and sale of forest products and about production for household use are divided between men and women. Women, for instance, were ignorant of decisions made about rights over resource use or payment of loans, while men were ignorant about decisions made by women concerning production for household use. Generally, the decisions made by women were unlikely to affect their legal or political access to resources. However, decisions made by men directly affected women's physical access to resources. Men often withheld information from women, whether deliberately or not, so that women were less able than men to control or influence decisions.

Not all women were equally dependent on forest resources to be able to provide for their needs: some wealthier women are able to use the resources from their private trees to meet household needs. Also in decision-making processes, their roles may differ: young unmarried women from wealthy households spoke out at a forest meeting, whereas young married women from the same households did not. Similarly, women who had attained the highest status within their households were also more assertive in a mixed group of women and men.

Source: Hobley (1991)

---

to the resources. These resources include land, water, equipment, labour, cash, leadership, representative organisations, education and information. Men and women also do not always have equal *control* over the resources, i.e. the ability to decide on their use. In West Africa, for example, the (male) head of the household decides on which fields his wives have to work. Women can only work on their own field after this assigned work is finished.

All of the differences described above are known as "gender differences". Box 1.4 gives a more elaborate example.

Because of all these differences, men's and women's views on needs and priorities to improve their situation often differ strongly. For instance, men and women within a household or a community may express a need to improve their living situation through a better food situation. Men may opt for fertiliser to increase cash-crop production to be able to get more cash and buy food, while women may give priority to a better water supply to grow more food crops. These needs may concern the day-to-day situation, but also more strategic issues to ensure improvements in the long term.

GENDER ANALYSIS

Although there are great differences between men's and women's roles in activities and decision-making, these roles are often strongly interrelated. Development efforts focusing on women will affect the men's position, and vice versa. It is a key challenge to PTD practitioners to take these interrelationships into account.

Drawing from several analytical frameworks published elsewhere, Box 1.5 gives a synthesis of the above considerations in the form of 10 critical areas of concern in developing a gender-sensitive approach to PTD. They form the basis for a so-called "gender analysis": a systematic effort to document and understand the tasks and responsibilities of men and women of different social and ethnic groups within a given context and to identify which groups will gain and which will lose from a proposed intervention, in order to minimise harm and/or maximise benefit to the disadvantaged in the process.

BOX 1-5 TEN AREAS OF CONCERN IN A GENDER ANALYSIS WITHIN PTD

1. Activities and responsibilities of men and women in PTD focusing on productive work but including attention to reproductive and community work (farmer groups).

2. Time allocated to different activities.

3. Access to relevant resources and to the benefits derived from using them.

4. Control over these resources and over the benefits from them.

5. Role and participation in decision-making processes at household and community level, including informal and formal organisations and informal decision-making.

6. Needs and priorities for improving the present situation, in both the short and long term.

7. Effects and impact of certain trends and developments on various groups, including economic (price fluctuations, subsidies), demographic (male migration), environmental (drought), socio-cultural (religious movements), political (changes in government, conflicts), and legal developments (ownership changes).

8. Constraints and barriers to active involvement in development efforts and ways to overcome these.

9. Opportunities and options to improve the situation: selection, benefits and impacts.

10. Organisations involved in local development and their capacity to deal with gender issues.

IMPLICATIONS FOR PTD PRACTICE

Once the importance of gender differentiation has been realised, numerous practical questions arise at various stages in the PTD process, such as:

- About whom and what should we seek information?
- From whom in the village should we ask for information?
- With whom should we establish contacts?
- Which critical items do we need to include in our diagnosis to become aware of possible gender issues?
- Who do we include in the diagnosis?
- Which diagnostic methods are we going to use to ensure involvement of different groups?
- Whose priorities are we going to address?
- Who should participate in experimentation?
- What are the implications of the proposed options for different groups?
- Who should be involved in evaluating the results?
- Who should be involved in farmer-to-farmer extension and training?
- How should we stimulate the formation of farmer-experimenter groups?

Problems may arise, however, in trying to use certain sources of information on these issues (see Box 1.6).

Finally, Boxes 1.7 and 1.8 give a number of practical guidelines to ensure that women farmers will be able to participate fully in PTD activities such as meetings, PRA events and training courses.

BOX 1-6 GATHERING INFORMATION ON GENDER ROLES FROM SECONDARY SOURCES

Statistical evidence on gender roles in agriculture is very unreliable. In many societies, it is culturally unacceptable both for a woman to say that she works in agriculture and for the census-taker to consider that she might have an economic role. Detailed fieldwork has often indicated a much higher level of female participation in agriculture than is generally recorded in national censuses.

Changes in employment status – e.g. from independent cultivator to unpaid family worker with the expansion of cash cropping in Africa, or from independent cultivator to wage labourer as landlessness increases in India – appear to have affected women workers more than men. Some of the variations in the recording of the role of women in national censuses may reflect societal changes in the perception of women's roles.

Source: Momsen (1991)

BOX 1-7 GIVING WOMEN FARMERS GREATER ACCESS TO MEETINGS AND COURSES

- Schedule meetings and demonstrations during women's free(est) time in the day or evening.

- Locate meetings and demonstrations where they are convenient for women to attend, e.g. close to homesteads.

- Locate demonstration plots along frequently travelled paths.

- Arrange meetings and other events at places accessible to and in ways appealing to women.

- Meet with women while they are pounding grain, or working communally or selling at the market.

- Hold courses at the time of year in which the women have least work.

- Provide transport to training/meeting centres.

- Provide separate residences for women at training centres.

- Provide childcare facilities or encourage cooperative childcare.

- Break courses into smaller modules, as it is easier for women to attend a two-day than a four-day course.

- Provide training in homesteads for women in seclusion.

- Ensure that the training content is interesting to women and relevant to their needs, roles and responsibilities.

Source: adapted from Saito & Spurling (1992)

BOX 1-8 GUIDELINES TO INCLUDE A GENDER PERSPECTIVE IN PRA ACTIVITIES

- Include male and female team members, to reach both women and men.

- Be sure that the team members are sensitive to gender issues and able to integrate a gender perspective into the analysis.

- Consult both women and men to determine a time, season and duration convenient for women and men to participate.

- Choose a location at which both women and men feel at ease.

- Select both male and female informants.

- Encourage women, as well as men, to express their own views.

- Consider forming all-female and all-male groups to ensure that both women and men can express their views and ideas.

- Start with a PRA activity that will ensure the interest and participation both of men and of women.

- Differentiate information in maps/matrices according to gender.

- Use a (local) language that both women and men can speak.

LEARNING ACTIVITY 1
ROLES AND BEHAVIOUR OF MEN AND WOMEN: DISCUSSING STEREOTYPES

1–1½ hrs

Objectives:
- To enable the participants to assess critically their own views on women and men farmers.
- To raise awareness of the implications of this for involving men and women in the participants' own work situation.

Setting/approach:
- Workshop setting in which small groups discuss widely heard statements on men and women.

Materials:
- Materials to present results of groupwork, such as newsprint sheets and markers; a handout with commonly heard statements on roles of men and women farmers (examples are given in the handout below).

Procedure:
- Distribute the handout with stereotypes about men and women farmers to the participants and ask them to read it individually and thoroughly.
- Ask the participants to form small groups of 3–5 persons, in which they discuss the stereotypes based on their own experience. Central questions for each statement are:
  - Does the group agree with this?
  - Why or why not?
- The main outcomes of the group discussions are presented and discussed in plenary, each group focusing on those statements that appeared to be most controversial in the group.
- Finally, discuss in the plenum the implications of these views on the way women and men farmers are reached in the participants' own work situation. The facilitator may need to pose some very critical questions in order to challenge participants' views.
- Ask all participants to mention two learning points from the discussions.

Variation:
- Statements in the handout given below need to be adapted and/or replaced to arrive at a collection of views most commonly heard locally. The facilitator may want to include similar sweeping statements made by participants themselves during previous sessions.

Source: Canadian Council for International Cooperation (1991); based on idea of Edith van Walsum, ETC (personal communication)

HANDOUT 1·2

OLD AND NEW STEREOTYPES OF WOMEN AND MEN FARMERS

1. Women do most of the agricultural work as unpaid farm workers, but are not involved in decision-making and do not control resources.
2. Men cultivate cash crops, while women cultivate food crops in small plots only for subsistence.
3. Women are more concerned with family welfare than with their own economic benefits and security.
4. Women can be reached by agricultural services and resources (e.g. extension, training, credit) indirectly, through their husbands.
5. Most families are composed of a man (head of household), a woman and their children.
6. Men are better with numbers and technology, while women have difficulty working with numbers.
7. Women farmers are poor and cannot afford to buy fertilisers, improved seeds and pesticides, and they are not able to make important agricultural investments.
8. Women do not respond to price and other incentives, since they are only subsistence farmers.
9. Women are overburdened with work and therefore cannot participate effectively in development activities.

LEARNING ACTIVITY 2
CASE STUDY: "THE BEAN EXPERIMENT"

1½–2 hrs

Objectives:
- To raise participants' awareness of the different priorities of men and women, based on their roles and responsibilities.
- To increase understanding of the different constraints to participation by men and women and of the different strategies needed to ensure participation of both.

Setting/approach:
- A workshop setting in which participants analyse in small groups a short case such as that given in the handout below.

Materials:
- Handout with case study and assignment for groupwork. Materials to present the results of the groupwork in the plenary session, such as newsprint sheets and markers.

Procedure:
- Distribute the case "The Bean Experiment" to the participants and ask them to read it.
- Let the participants form small groups of 3–5 persons, and ask them discuss the following questions.
  - How would you set up an experiment, e.g. on the different lines of bush beans, to be able to obtain results with data differentiating preferences according to gender?
  - Would women be able to take part in the experiment? Which kind of constraints do you envisage that might prevent women's participation?
  - What suggestions do you have to overcome such constraints in your own work situation?
- Systematise the results of the group discussions in a plenary session and discuss the outcome. It may be useful to present and discuss the above questions one by one. Pay attention to constraints based on factors within the household, the relation between husbands and wives, the project organisation and external factors.

HANDOUT 1.3

"THE BEAN EXPERIMENT"

Individual farmers were shown samples of seed from different lines of bush beans identified as promising for their area by the CIAT bean programme. Each farmer was asked to indicate those grain types that were of interest and those less acceptable. Their ranking varied somewhat from that of the breeders, because their most important criterion for acceptability was grain size. There was, however, one intriguing exception to this rule: the interest shown in a small-grain variety.

The interviews in which farmers made these initial selections were analysed. This suggested that the unexpectedly high ranking given to the small-grain variety was the result of women taking part in the selection. The women perceived that, traditionally, small-grain varieties similar in appearance to this type had been more flavourful and higher yielding. Women regarded a small-grain type as desirable from the viewpoint of the subsistence and consumption objectives of the small farm. Men, on the other hand, were selecting grain types for larger size, primarily with a view to marketability.

Source: based on Ashby et al. (1990)

LEARNING ACTIVITY 3
STRENGTHENING AN ORGANISATION TO DEAL WITH GENDER ISSUES

1½ hrs

Objectives:
- To obtain insight into methods and tools to increase the participation of women in agricultural development activities.

- To be aware of the capacity of one's own organisation to deal with gender issues.

Setting/approach:
- A workshop setting in which participants analyse in small groups a handout presenting some methods and tools. The need for specific attention to increasing the participation of women should have come out of previous sessions.

Materials:
- Handout 1.4 (p.74); materials to present the results of small groupwork in the plenary session, such as newsprint and markers.

Procedure:
- Distribute the handout and ask the participants to read it individually.

- Form small groups to discuss the different examples in the handout, using the following questions:

  • Which of the examples given do you think will be most effective? Do you have experience with these measures yourselves?

  • Which of the examples do you think will not be effective? Why not? Do you have experience with these measures yourselves?

  • What examples could you add that are effective in increasing the number of female beneficiaries?

- The results are presented and discussed in a plenary session. In this discussion, emphasis may be laid on the differences between measures relating to gender differentiation (points 1–4 in the handout) and those aimed specifically at facilitating the participation and involvement of women farmers (points 5–10).

- Conclude the plenary session with brainstorming on the following two questions:

  • What are the implications of this discussion in relation to the skills, knowledge and attitudes that the staff of our organisation should have in order to deal effectively with gender issues?

  • What realistic activities or measures (such as in training or female staff recruitment) should our organisation consider to improve skills and knowledge of the staff and to create a positive attitude towards gender issues?

HANDOUT 1·4 – LEARNING ACTIVITY 3

MEASURES TO INCREASE THE NUMBER OF FEMALE BENEFICIARIES

1. Change the focus of activities, e.g. increase the relative importance assigned to "women's" crops or animals.

2. Clarify specific gender aspects regarding roles, resources and project activities and their implications for project strategies, e.g. the implications of women's lack of access to land or men's lack of labour for weeding.

3. Examine the distribution of benefits, e.g. ensure that direct returns for women outweigh any additional effort.

4. Improve the messages, e.g. broaden the research agenda to cover the enterprises and tasks of women farmers, conduct more on-farm research on women's fields, appoint female subject-matter specialists.

5. Diversify information supply to include information, technologies and facilities that women specifically need, e.g. market information, appropriate tools and equipment, training in operating and maintaining equipment.

6. Adapt credit components, e.g. focus on smallholdings, reduce the minimum size of loans, use group liability rather than land title for collateral, reduce costs by loaning to groups.

7. Choose the appropriate language and communication network, e.g. use the vernacular, recruit local agents who speak the vernacular, use verbal or pictorial communication rather than written, communicate with groups rather than individuals.

8. Change the criteria for selecting contact farmers or for membership in groups.

9. Train male agents to work with female farmers.

10. Improve the location and timing of activities, e.g. provide one-day or mobile training rather than residential, hold evening meetings, identify a time and place where women congregate (markets, sites for communal work, etc) and use these settings as entry points.

11. Improve residential training for women farmers, e.g. provide childcare or separate boarding facilities for women.

Source: Saito & Spurling (1992)

UNIT 1·G / SKILLS IN COMMUNICATION AND PERCEPTION

OVERVIEW OF THIS UNIT

EXPECTED RESULTS

To be effective in their role as facilitator, advisor and supporter of farmers' efforts, PTD practitioners need to be able to communicate with farmers in true dialogue. Yet, this crucial skill is often neglected in formal education. Dialogical communication starts ultimately with realising how people select, project and interpret in their perception of reality, influenced by their cultural background, socio-economic position, education, etc. The challenge is to accept and respect these differences in perception, but still try to build bridges between them.

After completing the learning activities in this unit, participants are expected:

- to understand the basic principles of dialogical vs monological and mechanical communication;
- to have improved their skills to communicate dialogically;
- to be sensitive to the perception process and the role that differences in perception play in communication, especially in a cross-cultural context.

MAIN CONCEPTS

- **Dialogical communication:** Partners of equal status exchange views and information which are mutually understood.
- **Probing:** Continued questioning to unobtrusively stimulate the partner's flow of thought and analysis.
- **Non-verbal communication:** Facial expressions and inclination of the body are among factors that communicate opinions and interest without use of the spoken word.
- **Perception differences:** People observe and interpret the same reality differently because of personal as well as socio-cultural differences, their background and previous experiences.

TRAINING METHODOLOGY

As the topics of this unit are very close to the participants' personal development, the learning activities should take place within the "safe" workshop setting. Much existing literature on communication suggests possible activities in this field; and trainers will often have their own set which they normally use. Several examples are included in this unit to stimulate creativity. Further ideas can be obtained from the sources on participatory training listed in the Appendix to this book (p. 223).

LEARNING ACTIVITIES

1. Role playing: dialogue and monologue
2. Listening pairs
3. Formulating open questions
4. Formulating probing questions
5. Perception exercises

DISCUSSION

DIALOGICAL COMMUNICATION

PTD requires a dialogical communication, in which two partners of equal status can exchange their views by means of messages which are mutually understood. Dialogical communication stimulates common action. The contrast is monological communication, in which one party (the "sender") determines the messages without giving the other party (the "receiver") any chance to express his or her ideas and comments. As a result, monological communication reduces one party to passivity and, in the long run, produces a frustrating situation in which much human potential is lost. PTD practitioners are aware of the value of dialogical communication and its consequences for PTD work, especially in interaction with farmers who have previously been forced into the role of passive receivers.

This unit highlights several important concepts and skills of such dialogical communication.*

LISTENING WITH AN OPEN MIND

PTD practitioners aim at communicating receptivity and respect and at listening to what the farmers are saying with an open mind. Certain habits prevent listening well:

- after hearing parts of a story, thinking that we understand the main points and letting our minds wander to other matters;
- becoming upset or angry at certain words or phrases, and ceasing to listen;

---

* This section draws mainly on overviews made in Ashby (1990) and Hope *et al.* (1984)

- quickly feeling that what we hear is boring or does not make sense;
- daydreaming;
- if we do not immediately understand what we are listening to, letting our minds wander or close, rather than asking for clarification;
- if we hear something that seems to challenge some of our favourite ideas and prejudices, rejecting it inwardly.

A long list of dos and don'ts in listening can be put together. Such a list can be developed in a training session together with participants, by working through different examples and activities. Box 1.9 summarises some important "Dos". Unit 2.C (pp.113–28) includes further details on listening in interview situations.

PROBING

Probing is the natural sister of listening. It aims to combine receptive listening with questions that unobtrusively direct the dialogue partner's flow of comment. At the same time, it provides a crosscheck on one's understanding of the other person's point of view and the consistency of her/his remarks. Apart from the questions mentioned in Box 1.9, probing techniques used in dialogue with a farmer may include those in Box 1.10.

In a discussion with farmers, especially when trying to learn farmers' views and elicit farmers' knowledge, one may ask three types of questions:

- **Leading questions:** The speaker tries to get the farmer to agree with the speaker's viewpoint ("Don't you think that..?"). Leading questions are all too common in everyday conversation, but should not be used in PTD discussions;
- **Direct questions:** In discussions with farmers, direct questions may be used if the aim is to obtain specific points of information, e.g. How often? How much? Which variety?
- **Open questions:** These do not direct the response of the dialogue partner and allow free expression.

BOX 1-9 LISTENING TECHNIQUES

TYPE	PURPOSE	POSSIBLE RESPONSES
Clarifying	To get at additional facts. To help the person explore all sides of a problem.	Can you clarify this? Do you mean this? Is this the problem as you see it now?
Restatement	To check our meaning and interpretation with the other person. To show you are listening and that you understand what the other person has said.	As I understand it, your plan is ... Is this what you have decided to do..? And the reasons are ..?
Neutral	To convey that you are interested and listening. To encourage the person to continue talking.	I see. I understand. That's a good point.
Reflective	To show that you understand how the other person feels about what s/he is saying. To help the person evaluate and temper his/her own feelings as expressed by someone else.	You feel that ... It was shocking as you saw it. You felt you didn't get a fair hearing.
Summarising	To bring all the discussion into focus in terms of a summary. To serve as a springboard to discussion of new aspects of the problem.	These are the key ideas you have expressed. If I understand how you feel about the situation...

> **BOX 1·10 EXAMPLES OF PROBING QUESTIONS IN DIALOGUE WITH A FARMER**
>
> - Restate what the farmer has just said (the mirror technique): *"So it resists the drought …"*.
>
> - Repeat a remark that has just been made, in the form of a question. By doing this, you invite the farmer to expand on this particular theme: *"It resists the drought?"*.
>
> - Go back to and repeat a comment made earlier. This can help to steer the farmer's flow of comments in a direction you think important: *"Earlier you said…"*.
>
> - Ask the farmer to clarify: *"Could you tell me a bit more about this?"*.
>
> - Summarise in your own words what you understand the farmer has said and ask: *"Do I understand correctly?"*.
>
> - Be prepared to admit uncertainty with the statement: *"I'm not sure I understand correctly; you seem to be saying …"* and repeat the farmer's statement.
>
> - Remain silent (the five-second pause), maintaining eye contact. This encourages the speaker to keep talking.

Open questions form the main element in PTD discussions. Examples are given in Box 1.11.

Such open questioning becomes stronger if reasons for the foregoing statements are sought, as in the "But-Why Method":

Farmer: *"Yields of corn are generally low here."*

PTD practitioner: *"Why?"*

F: *"Because the rain stops before the crop matures."*

P: *"But why?"*

F: *"Because we always plant late."*

P: *"But why do you plant so late?"*

F: *"Because we don't have any time earlier."*

P: *"Why?"*

F: *"Because we first have to help plant the chief's field."*

P: *"But why do you have to?"*

F: *"Because we are in debt to him."*

… and so on.

Even if the local culture does not permit the use of the direct "But why" form, the same idea can be put across using other words. It is through asking such questions that real learning can take place for all involved.

BODY LANGUAGE AND NON-VERBAL COMMUNICATION

A fieldworker, consciously or unconsciously, communicates a great deal to farmers without actually saying anything. Facial expression, inclination of the body, where s/he sits in relation to the farmers: these and many other forms of body language and non-verbal communication indicate to the farmers how keen and honest the fieldworker's interest is in their situation. The example in Box 1.12 (p. 78) illustrates how a fieldworker may stress his/her imagined superiority by standing above a sitting farmer, or at some distance from the farmer, but can also consciously change this position. PTD practitioners are aware of how things are communicated without words and make positive use of the possibilities.

PERCEPTION

Dialogical communication between people from different cultural backgrounds is possible only if each realises that the other perceives reality differently. There is no objective reality, only many subjectively perceived realities. This subjective perception guides our behaviour, including our communication with others. Realising this should make it easier to understand and respect other cultures.

> **BOX 1·11 OPEN QUESTIONS TO STIMULATE FARMERS' IDEAS**
>
> - Can you tell me more about this?
> - What would be an example of that?
> - What makes you see it this way?
> - What are some reasons for that?
> - Could you help me to understand this better?
> - Have you any other ideas about this?
> - How do you feel about that?
> - How do you think other farmers would feel about this?
> - How would you describe this?

Perception has a personal but also a cultural dimension, since thinking, feeling and behaviour are learned within groups: the family, the social group or the larger community. This helps explain some of the difficulties that arise in intercultural communication between farmers and fieldworkers. Their cultural background may be very different: rural vs urban, poor vs relatively rich, female vs male, powerless vs relatively powerful, old vs young, illiterate vs literate, etc.

When we look at the world, we are all selecting, projecting, interpreting and attaching meaning to what we see (Payr & Sulzer, 1981):

- **selection** means that, from everything perceived by our eyes, ears and nose, we choose only that which interests us, only what we actually need to use at that moment, while neglecting everything else;

- **projection** indicates how our existing feelings, fears or wishes influence and colour what we perceive;

- **interpretation:** whether we realise it or not, we "store" our perception only after we have given it a certain meaning, have ordered it on the basis of what we knew already, so that it makes sense and we can remember it later, when needed.

Unfortunately, the modern education of fieldworkers – especially, but not only, those who are trained in agriculture – has strongly influenced the way they perceive agriculture. Instead of farmers' holistic and often spiritual perception, scientifically trained staff have been taught to regard crops, livestock, soil, water, etc as isolated parts of the process, which can be understood only by studying each individually and in ever greater detail, rather than seeing them within a web of relationships and meanings. Therefore, many fieldworkers need to become aware of how they perceive things and, as far as possible, to learn new ways of perception.

BOX 1-12 COMMUNICATION WITHOUT WORDS: 3 EXAMPLES

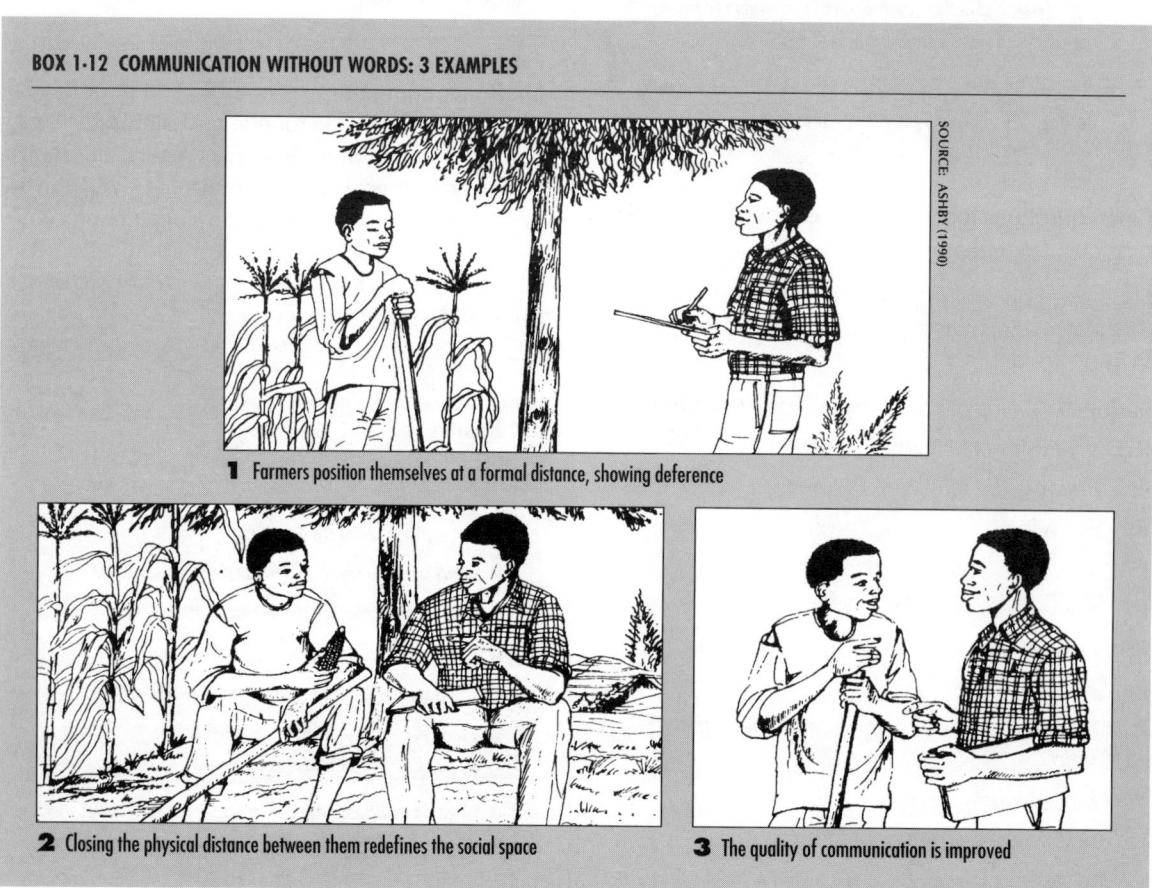

1 Farmers position themselves at a formal distance, showing deference

2 Closing the physical distance between them redefines the social space

3 The quality of communication is improved

LEARNING ACTIVITY 1
ROLE PLAYING: DIALOGUE AND MONOLOGUE

45 min

Objective:
- To increase participants' awareness of the value of listening, to allow dialogue rather than monologue.

Setting/approach:
- Short role play by six participants. As it is relatively simple, it may be used early in a workshop.

Materials:
- Pen and paper, newsprint sheets and markers for the plenary discussion, and possibly attributes to make the play more convincing.

Procedure:
- Invite six people to prepare a short play in three scenes. It is usually better to have all women or all men acting, as this avoids people saying, "men always do this…" or "women always do that…".

SCENE 1:
Two people meet. One of them starts to talk and gets so excited and involved in what s/he is saying that s/he pays no attention to the other. The other person tries several times to speak, ask a question, respond or make a suggestion, but the first person talks on, so the second remains silent and gives up trying. (The pair should decide on a topic beforehand.)

SCENE 2:
Two people meet and each starts telling the other what s/he is concerned about. They are each talking about a different topic. Neither is listening to the other, and both are talking at the same time.

SCENE 3:
Two people meet, greet each other and begin a real dialogue. Each asks questions about the other's interests, listens and responds to the other's answers, and shares her/his own news and opinions. A common topic should be decided on beforehand.

- One of the trainers should stop each play as soon as the point has been made. Usually, the first two plays take 1–2 minutes and the third play takes a little longer.

- Divide into 3-person subgroups to answer the following questions:
  • What did you see happening in Scene 1?
  • What did you see happening in Scene 2?
  • What did you see happening in Scene 3?
  • Do these things happen in real life? For example?
  • What can we do to help make communication as good as possible in this group?

- The whole group comes together and verbally shares briefly their answers to the first two questions. The trainer asks someone to write on newsprint the answers to the last question only.

- The trainer summarises the points on the last question. Keep the newsprint sheets on the wall to give the group its own "Guidelines for Good Communication".

Variation:
- A similar activity can be developed to discuss non-verbal communication and body language. Participants may be asked, in small groups, to select typical examples of fieldworkers' behaviour that puts farmers ill at ease and to develop a role play around it. The group may than identify ways to improve communication and prepare a short play around this too. Male/female mixes in such groups often come up with interesting cases. Showing the plays to each other provides a basis for joint understanding of non-verbal communication.

Source: Hope et al. (1984); variation based on ideas from Louise Sperling (pers. comm.)

LEARNING ACTIVITY 2
LISTENING PAIRS

about 1 hr

Objective:
- To enhance participants' skills in careful listening.

Setting/approach:
- Discussion exercise in pairs, to be used after group members have become fairly well acquainted with each other.

Materials:
- Newsprint sheets, markers, and tape for plenary discussion.

Procedure:
- Each person is asked to find a partner with whom they know they disagree on a specific subject. The pairs are then asked to discuss this subject, but after each one speaks, the other must summarise to the speaker's satisfaction what has just been said, before giving his or her own response or point of view.

  Note: In this exercise, the pairs choose for themselves the topic they will discuss.

- ALTERNATIVELY: Each person is asked to choose a partner and the trainer gives a controversial topic for them to discuss. Same procedure as above.

  Note: Apart from issues directly related to the content of the course, possible topics are: family planning, divorce, women's liberation, socialism/capitalism, smoking rules in the group.

- After the above exercise, the trainer asks the group what difficulties they experienced in listening and lists these on newsprint.

- The trainer asks what they can do to improve communication in the group. The answers are written on newsprint.

Variations:
- Instead of working in pairs, the participants may be asked to work in groups of three. In each group, two participants discuss a topic as described above. The third member of the group acts as referee, checking whether the summaries given truly reflect the ideas of the previous speaker. The procedure in the small groups would then be: Participant A gives his/her view on the matter at hand; Participant B summarises the view expressed by A before giving his/her reactions; the referee (Participant C) briefly checks with A whether the summary given truly reflects his/her words. B then proceeds to give his/her reactions, A summarises, C checks this summary, and so on. This procedure should be shown once in plenum by the trainers to clarify the different steps.

Source: Hope et al. (1984)

LEARNING ACTIVITY 3
FORMULATING OPEN QUESTIONS

45 min

Objective:
- To enhance participants' skills in formulating open questions to stimulate the spontaneous expression of opinions by farmers.

Setting/approach:
- An individual exercise followed by group discussion.

Materials:
- Handout "Interviewers' Questions", pens, blackboard.

Procedure:
- The theme is introduced and one or more examples of open-ended and leading questions are given.
- Handout 1.5 is distributed to participants, who are asked to reword the questions given in the left-hand column into an open-ended format.
- The individual answers are then analysed by the whole group.

Variation:
- Ask one participant in the group to formulate other leading questions and another to reformulate these as open-ended questions; the other participants may comment on the rewording.

Source: Quiros et al. (1990)

HANDOUT 1.5

INTERVIEWERS' QUESTIONS: Individual Assignment	
LEADING QUESTIONS	**OPEN-ENDED QUESTIONS FOR PROBING**
I suppose you think it's good because of the number of maize cobs?	
Don't you think the plants are too short?	
Because of the colour, right?	
By yield, you must mean it has more pods per plant. Isn't that right?	

LEARNING ACTIVITY 4
FORMULATING PROBING QUESTIONS

45 min

Objective:
- To enhance participants' ability to pose probing questions that stimulate the farmer to explain an opinion in an evaluation.

Setting/approach:
- Individual exercise followed by group discussion.

Materials:
- Handout "Farmer's Comments", pens.

Procedure:
- Introduce the topic and give more examples of probing by briefly interviewing a participant.
- Distribute handout 1.6 and ask participants to write, after each of the farmer's comments, a probing question that would stimulate him/her to explain his/her opinion in greater detail.

Source: Quiros et al. (1990)

HANDOUT 1.6

	FARMER'S COMMENTS: Individual Assignment
	F = farmer; I = interviewer
1.	**F** The plant is well developed, is full of leaves and has many pods.
	I
2.	**F** You see, it's really leafy; I mean well developed or bushy, and so it will produce more; the leaves aren't too big or too small.
	I
3.	**F** What I always look at is the number of seeds the pod has, and this one has 4–5 in each.
	I
4.	**F** The thickness of the grain, because then it goes further in filling the sack.
	I
5.	**F** I like to grow it for selling it, as the price is good; of course, I always keep a little for the house.
	I
6.	**F** I think so; it's streaked, like the variety Calima, which everyone asks for. It's the most popular one.
	I
7.	**F** Well, what should I say … Uh, look, I think this is a little diseased.
	I

LEARNING ACTIVITY 5
PERCEPTION EXERCISES

30–60 min

Objectives:
- To increase participants' awareness of the perception process and distortions that occur because of differences in cultural background, education, socio-economic factors, etc.
- To enhance awareness in participants' own ways of selecting, projecting and interpreting when observing and studying "reality".

Setting/approach:
- Individual assignment in a workshop setting.

Materials:
- Objects, photographs or drawings (like the one shown below) for observation.

Possible procedures:
- The basic feature of the following perception exercises is that participants are asked to observe and interpret a situation which makes evident the subjective character of perception.

Examples:
- An object or abstract drawing is shown briefly to everybody. It is then described on paper by two participants, and the results are compared. To a large extent, the differences between the descriptions will indicate differences in perception.

- Some photographs of different cultural settings are shown to the group or to individuals; the participants are asked to note what they learned from the photographs. Comparison of the results will generate a discussion on perception processes.

- Participants may also be invited to add examples from their own experience while working with farmers and to try to give an explanation according to mechanisms of human perception.

Source: Svendsen, D. & Wijetilleke, S. (1983)

A family with problems... What do you see? Where to go for proper assistance?

TOWARDS AN AGENDA FOR ACTION

PART 2

PART 2 / TOWARDS AN AGENDA FOR ACTION

2·A / **GETTING STARTED**	■ Where to start	90
	■ Existing information and documentation	90
	■ Organisational inventory	92
	■ The PTD team	92
	■ Establishing rapport	93
2·B / **PARTICIPATORY SITUATION ANALYSIS: STARTING POINTS**	■ The need for participatory analysis	102
	■ Communities are not homogeneous	102
	■ Problems and opportunities	103
	■ Preventing biases	104
	■ Farmer-led analysis	104
2·C / **METHODS IN PARTICIPATORY SITUATION ANALYSIS**	■ Central features	114
	■ Interaction methods	115
	■ Tools and Techniques	117
	■ Selecting and combining methods, tools and techniques	121
2·D / **LOOKING FOR THINGS TO TRY**	■ Maintaining a participatory mode	129
	■ Generating farmers' options	130
	■ Linking with ideas elsewhere	131
	■ Farmers' and outsiders' criteria	131
	■ Making a choice	132
	■ Hypotheses for research	132

BOXES	2.1	Criteria to select the location of programme activities: an example from Indonesia	91
	2.2	Example of a simulation play: "Village Unknown"	101
	2.3	Answers to the questions of the Zambia case	110
	2.4	Problems of a bus company: a problem tree	112
	2.5	The role of visualisation	114
	2.6	Users' notes on semi-structured interviewing	116
	2.7	A bioresource flow diagram	117
	2.8	A transect	118
	2.9	Gender-differentiated seasonal calendar	119
	2.10	An example of direct matrix ranking	120
	2.11	Women farmers choose bean seeds for trials	131
	2.12	Hypotheses for a farmer-led trial on zero grazing	132
	2.13	An idea sheet	133
TABLES	2.1	The source and use of secondary information	91
	2.2	Farmer differentiation matrix	103
HANDOUTS	2.1	Average monthly rainfall for Middle Kirinyaga	97
	2.2	Maize and bean prices in Middle Kirinyaga, 1979–81	97
	2.3	Is this a problem?	109
	2.4	Maize production in Kabwe District, Zambia	111

UNIT 2·A / GETTING STARTED

OVERVIEW OF THIS UNIT

EXPECTED RESULTS

At the start of any new PTD programme a number of issues need careful consideration. These issues include: which area to work in; who selects these areas; how to prepare oneself for a PTD programme; and how to establish the first contacts in the field.

Having gone through the learning activities of this unit, fieldworkers are expected to be aware of the key issues in preparing for PTD programmes and be able to apply these in their work. More specifically, fieldworkers are expected:

- to be able to reflect critically on their own criteria for selecting work areas and other actors' (farmers') criteria for choosing work areas;
- to realise the importance of giving farmers or communities the initiative in PTD as soon as possible;
- to be aware of, and consider, existing information before entering an area;
- to be able to identify, assess and involve other organisations as possible partners in, or contributors to, the PTD process;
- to be aware of important issues in the first contacts with villages and farmers and to be able to initiate such contacts.

MAIN CONCEPTS

- **Farmer or village self-selection:** Initiatives for activities are in the hands of farmers as soon as possible.
- **Three categories of selection criteria:** Felt needs and the existence of problems; awareness and willingness to act upon the problems; and the agro-ecological and socio-political potential.
- **Organisational inventory:** Determining who is doing what in the work area which is relevant for the PTD process.
- **RAAKS (Rapid Appraisal of Agricultural Knowledge Systems):** Elaborated approach to involve all relevant organisations in analysing problems or issues at hand.
- **Developing rapport:** Building of a relationship of trust and of joint commitment with possible collaborating organisations in the PTD programme.

TRAINING METHODOLOGY

Many of the basic insights, attitudes and skills relevant here have been dealt with in Part 1; for example:

- understanding and respecting rural life (Unit 1.E, pp. 59–66),
- village organisational structures (Unit 1.E, pp. 59–66),
- gender aspects (Unit 1.F, pp. 67–74), and
- cross-cultural communication (Unit 1.G, pp. 75–83).

It is important to integrate these issues into this unit's learning activities (pp. 94–101). They will be highlighted either by brief role plays and discussions or by simulating first contacts with individual farmers and/or more elaborate role plays and discussions.

Several specific issues can be addressed by asking participants to reflect on their present practices, either in a brainstorming session, where all participants are asked to share their experiences, or by case studies of the experiences of selected participants. Existing data and maps can be made available during the workshop for participants to practise selective extraction of relevant information. Participants can also be asked to present to other participants in a concise way what they concluded from the secondary data. They may also be asked to prepare an organisational inventory for the area in which they work, or for the area in which they will carry out a diagnosis during the workshop.

Practical fieldwork in one or more villages during the workshop gives, of course, a much more realistic learning situation. There are, however, drawbacks in this approach, as unprepared participants may make painful mistakes, for them or the villagers. Secondly, through the workshop a process of mobilisation will be initiated in the villages concerned, which may not receive follow-up after the training has finished.

LEARNING ACTIVITIES

1. Village selection
2. Brainstorming existing information sources
3. Analysing existing information
4. Role plays of first village contacts
5. Explaining PTD
6. Simulation: "Village Unknown"

See also the learning activities of Units 1.E, F and G.

DISCUSSION

WHERE TO START?

Many organisations wishing to initiate PTD activities have a set geographical area with existing contacts in villages. In other situations the choice of area or villages to work in is still open. Many development organisations then opt to select an area to work in, enter the villages and then push for their package of assistance. This practice puts the community from the start in a "receiving" position: "If you are so eager to start something in our village, that's fine, but we will wait and see."

The participation process gets a much better start in cases where the request for assistance comes from the village itself. To achieve this, a development organisation may introduce its approach in different areas or villages and explain where it comes from and how it works, but thereafter wait for official requests from the village (or villages) to start a collaboration programme. It should not be forgotten, of course, that such requests may represent the interests of a few individuals who may not necessarily belong to the target group of the programme.

Each organisation, however, also needs to define clearly its own criteria, if only to be able to judge requests for assistance coming in, or to select a region as an area of work in which to concentrate its efforts. This would include a clear understanding of which farmers are to be given priority in the activities envisaged, e.g. small or larger farmers, men and/or women, landowners or tenants, pastoralists or settled farmers. Box 2.1 gives a list of criteria used by an Indonesian NGO.

In general, three categories of criteria can be distinguished:

a. **the need:** there are severe problems felt by specific groups;

b. **the awareness:** the people are, to a certain extent, aware of their problems and willing to act; and

c. **the potential:** the situation makes a PTD-like process possible (internal cooperation, ecological potential).

On the other hand, there are often less official reasons to work in a certain area or village: e.g. political reasons when a government asks an organisation to work in an area to prevent unrest, or strong affiliations between NGO leaders and (part of) the population in the particular area.

EXISTING INFORMATION AND DOCUMENTATION

In almost all cases, a considerable amount of existing information on a particular area is available. Screening and studying this information may greatly increase the effectiveness of the PTD team at a later stage in the programme.

There are many sources for secondary data related to agricultural development. Table 2.1 lists some sources and the type of information they provide. However, caution must be taken – there are at least three distinct dangers or pitfalls in using these secondary data:

- The large amount of statistical data and detailed maps may draw too much of the attention to a quantitative analysis. Initially, one should be selective and use only those data that may directly contribute to understanding the main aspects of the farmers' situation in the area. More detailed information, determining, for example, the possibilities of certain new varieties, may be accessed and used at a later stage.

BOX 2·1 CRITERIA TO SELECT THE LOCATION OF PROGRAMME ACTIVITIES: AN EXAMPLE FROM INDONESIA

1. Relative poverty, presence of marginal groups

2. No other development programme active in the area

3. Accessibility (by foot)

4. Agricultural potential:
   - stable land tenure
   - relative importance of rain-fed agriculture
   - absence of free-roaming cattle

5. Relative openness to new ideas

6. Socio-cultural potential:
   - absence of major internal conflicts
   - supportive local leadership

7. Expectations of villagers towards development activities consistent with Propelmas approach

Source: Propelmas (unpublished)

TABLE 2·1
THE SOURCE AND USE OF SECONDARY INFORMATION

TYPE	SOURCE	ASPECTS FOR FARMING SYSTEMS DEVELOPMENT
Administrative map	Survey Department	Location of states, regions, districts, villages, delivery points of support services
Road maps	Survey Department, petrol stations	Airports, harbours, roads, tracks, railway connections, petrol stations, hotels, other accommodation, churches, mosques, tourist attractions, etc
Topographical maps	Survey Department	Topography, watersheds, rivers, reserved areas, forest
Soil and soil suitability maps	Soil Survey Department	Major soil types, suitability of major soil types, soil fertility, erosion and land degradation risk areas, etc
Land use maps	Survey Department	Land use information, urban, industrial, agricultural, cropped areas, livestock areas, sometimes containing quantitative information about planted areas, densities, etc
Aerial survey photographs	Survey Department	Updating land use patterns, plot size, detection of recent settlement areas, recent deforestation, changes in river courses, etc
Census reports	Statistics Office	Population data, structural and production data, usually covers many different subjects
Climatic data	Meteorological Department stations	Rainfall, temperatures, wind, hours of sunshine, evaporation, determination of growing season, crop risks, drought, floods, frost, hail, variability, intensity, etc
Macro-economic framework	Ministry of Finance & Planning, Ministry of Trade, universities	Macro-economic data, planning rules and procedures, foreign exchange regulations, internal and external trade regulations, inflation, price regulations, taxes, levies, exports, imports, quota regulations, domestic demand, etc
Experimental results	Experimental stations, research institutes	Suitable varieties for specific conditions, cultivation practices, pest management practices, yield response curves, post-harvest techniques, etc. Annual crops and perennial crops. Same for livestock.
Crop production	Min. of Agriculture	Planted areas, cost of production, yields, total production figures, etc
Irrigation data	Min. of Agriculture	Irrigated areas, type of irrigation, cost recovery policies, investment costs, feasibility studies, etc
Nutrition	Nutrition institute	Nutritional requirements, food preferences, identification of problem areas, food balance sheets, related health problems, etc
Health information	Ministry of Health, local clinics	Major diseases, number of doctors, hospitals, general health issues, etc
Anthropological, sociological information	Universities, research institutes	Traditions, customs, attitudes, social structure, local leadership organisation, role patterns in society, behaviour, etc

- The reliability of the data is often questionable, because of the way they were collected and processed. Conclusions emerging from studying these data, especially those related to agricultural problems, should therefore be treated carefully, as first hypotheses for confirmation from other sources, most notably the people directly involved.
- Many of the data on people and farmers show little differentiation according to, for example, economic status or gender. Recently, studies have become available with agricultural information segregated according to gender. Relatively new environmental studies or profiles may also be available.

Even if there are no documented data on the area, knowledgeable people may be found within the government services or the private sector with experience in the area. Focused interviews with such resource persons can provide the PTD team with valuable information before going to the field.

ORGANISATIONAL INVENTORY

In general, involvement and support from different organisations is required for successful PTD implementation, e.g. research organisations, extension organisations, NGOs, farmers' unions, credit organisations, land reform departments and local government. Enlisting their cooperation from the start increases the chances of sustaining the PTD process after the role of the PTD team "fades out".

One easy way to do this is to prepare an "organisational inventory" (Jiggins & de Zeeuw, 1992), which is basically a list of all the organisations working in the area with information on their programmes and activities, an assessment of their possible role in a PTD development programme, and an evaluation of its present impact, strong and weak points. A much more interactive and profound approach has recently become known under the name of RAAKS (Rapid Appraisal of Agricultural Knowledge Systems; Engel & Salomon, 1996). In the RAAKS approach all organisations, formal or informal, are challenged to give their views on the issues at stake; for example, agricultural problems and opportunities in the selected area. Major areas of concern are what the organisations are doing, or should be doing, to address these problems. These different views are brought together to develop a common understanding of what needs to be done to make the efforts of all involved more effective. RAAKS gives a number of "windows" or "fields of analysis" to help the PTD team facilitate the above analysis. There are three phases in RAAKS:

1. Problem definition within the team, carried out in a mini-workshop.
2. Constraint and opportunity analysis with actors in a workshop and individually.
3. Intervention planning.

Tools and techniques have been developed for use in each of these phases.

The RAAKS approach seems to be relevant especially when a considerable number of organisations are directly involved in agricultural development work in a particular area, and when the lack of collaboration and communication causes differences in views on what needs to be done.

THE PTD TEAM

In the interactions between outside agencies and the villagers, small "PTD teams" will often play a crucial role. Forming and training these is an important step in getting started. To enable effective interaction, these teams are generally small (1–4 members). There are no fixed rules for team size and composition as:

- these depend of the local situation, the agricultural issues at hand, and available resources;
- these vary over time; a first visit may be made by only one staff member, a second PRA-type visit may expand the team to three or four, while subsequent monitoring visits may need the involvement of only one or two.

In general, important questions to be asked in forming teams are:

- Are both facilitation skills and agricultural content expertise sufficiently represented?
- Are the most relevant disciplines brought in, technical, economic and socio-cultural?
- Does team composition enable gender issues to be seriously addressed? Is there a gender awareness in all members? Is there a gender balance in the team? Are female team members available to interview women?

Rapport and trust must be established between farmers and PTDers as a basis for collaboration in a participatory programme

ESTABLISHING RAPPORT

Before actual joint work can begin, farmers and outsiders have to start knowing and trusting each other. The community may not trust outsiders because of earlier experiences. Outsiders often do not know the community and may not trust the motivation of, for example, certain village officials. Trust-building continues through the PTD process, but must start at the very beginning (van Veldhuizen, 1990).

The crucial factor in this process is the attitude of the fieldworkers towards the community, and their skills in communication. Important questions are: Is there respect for farmers' way of life and their culture? Is there sensitivity to aspects of gender? Are they aware of the pitfalls of cross-cultural communication? (See Units 1.E, F and G, pp. 59–83.)

First activities therefore often take the form of explorative PRAs (see Unit 2.C, pp. 113–28). Parties to be met during these first contacts include:

1. the group or individual who initiated the request for assistance; this may be a village leader, but also an existing farmers' group, or individual farmers;

2. the formal and informal leaders of the village who may include:
    - government officials, both administrative and technical staff;
    - military or police officials, as sometimes their official approval is needed;
    - religious leaders, who are often very influential;
    - leaders of traditional institutions, chiefs, clan-heads, etc;
    - leaders of "modern" organisations, e.g. cooperatives and village development committees;
    - other informal leaders, who may not be part of any formal institution or structure, e.g. landlords, merchants, informal women leaders and traditional birth attendants;

3. the wider community, both men and women.

In practice, animosity may exist between different leaders, in which case the fieldworker has to be very sensitive as even the sequence in which s/he visits them may hinder the trust-building process.

Apart from individual meetings with key people, a community meeting should be seriously considered to emphasise from the beginning the wish to involve all village members/representatives in the envisaged activities.

Issues to be discussed during these first contacts will include:

- information from outsiders, their organisation, where it comes from and its affiliations with others (churches, unions);
- the explanation of the type of cooperation it is aiming at; some basic ideas underlying PTD;
- a clarification of the expectations of those who asked for assistance;
- general advice from village leaders on how to work in the village; and
- an outline of first activities after this initial visit: e.g. diagnostic activities, PRAs, etc.

Although, after these first contacts, the full implications of the envisaged PTD programme will not be clear to all involved, it is important that at this stage the community or group is asked to confirm explicitly its commitment to the programme on the basis of the previous discussions. In this way it becomes *their* programme. It may also be appropriate to formalise this decision in a simple written agreement between the programme and the farmers/community.

LEARNING ACTIVITY 1
VILLAGE SELECTION

1–1½ hrs

Objectives:
- To understand one's own criteria and other important criteria for area/village selection.

- To realise the importance of a community's own initiative in getting started and the consequences this has for the selection process.

Setting/approach:
- This is a brainstorming activity carried out in pairs, with a subsequent plenary discussion.

Materials:
- Pen and paper; for plenary presentation and discussion, cards, markers, and board to hang cards on.

Procedure:
- Ask the participants to form teams of two, remaining seated, and present to the teams two questions to be answered:

  1. What are important reasons for including or excluding a village from your programme?

  2. To what extent does the village itself influence the decision-making to initiate activities?

- The teams study these questions (15–20 minutes) and prepare for the plenary discussion by writing each reason for question 1 on a separate card, while preparing a short statement on the second question.

- In the plenary session, ask one team to hang their cards on the board and to provide a brief explanation; subsequent teams then place cards with similar issues close to those of previous teams, while cards with new issues are given a different position.

- Provoke reflection on the results of the brainstorming by asking:

  • What is the central theme of the group of cards that have been placed close together on the board? These may be summarised under headings such as "need", "awareness", "potential".

  • Are there controversial criteria? if so, let the teams explain their rationale for (not) including them.

  • Do the participants identify specific target groups within villages (rich/poor, men/women)? and how does this influence their criteria?

- Ask the teams to present the answer orally to the second question; and then provoke further reflection by asking:

  • In what ways can the role of the community in the selection process be increased? Is it realistic to expect initiative for activities from the community? Why is this important?

LEARNING ACTIVITY 2
BRAINSTORMING EXISTING INFORMATION SOURCES

45 min

Objective:
- To be aware of existing information and secondary data and be able to use it critically, before entering an area.

Setting/approach:
- This activity involves brainstorming sessions in small discussion groups.

Materials:
- Flipcharts, marking pens, Table 2.1 from the discussion section of this unit (p. 91).

Procedure:
- Ask the participants to form small discussion groups of a maximum of 4 participants.

- Ask the small groups to make a list on a flipchart with types of data available for their area, mentioning the source for each.

- (As an option) distribute after 15–20 minutes a handout based on Table 2.1 and ask the groups to use it to check and possibly improve their list.

- The subgroups are asked to present their results from the flipcharts briefly to the others.

- The plenary discussion could focus on:

  - How detailed do such data have to be, in order to be useful at the initial stages of starting PTD?

  - What do we know about the reliability of the data?

  - What can you learn from such data about different groups of farmers and their problems?

  - To what extent do the data have to be supplemented by field diagnosis in the area? What additional information will further diagnosis provide?

- A possible concluding question to the plenary discussion is:

  - To what extent are you using existing information in your present work? Is there too much reliance on such information or is it neglected? Why?

Variation:
- A similar brainstorming session is held, asking for relevant local organisations to be involved in future PTD programmes with their respective roles and interests.

Source: adapted from Walecka et al. (1987)

LEARNING ACTIVITY 3
ANALYSING EXISTING INFORMATION

2 hrs

Objectives:
- To be able to use existing information to develop hypotheses about farmers' problems.

- To enhance awareness on the need to differentiate between different groups of farmers.

Setting/approach:
- This is a case study exercise involving small work groups.

Materials:
- Handout 2.1, "Average Monthly Rainfall for Middle Kirinyaga"; Handout 2.2, "Maize and Bean Prices in Middle Kirinyaga, 1979–81"; blackboard and chalk, or newsprint and markers.

Procedure:
- Distribute the handouts "Average Monthly Rainfall for Middle Kirinyaga" and "Maize and Bean Prices in Middle Kirinyaga, 1979–81". Note that the price chart also includes planting and harvesting dates.

- Further explain that the rainfall and market price data used in this exercise are from the Kirinyaga District, Kenya. The area is at an altitude of 1200 metres, the topography is flat to mildly sloping, and the soils are deep loams. The area is inhabited by limited-resource farmers, farming about 2 to 5 hectares per family. The principal crops grown are maize and beans, which are also the most important food staples.

- Ask the participants to form small groups and then:

    1. Request that they prepare a list of hypotheses (nothing more than first ideas for further checking) about farmer problems, based on the information in the handouts.

    2. After the groups have worked on this for some time, give the following suggestions to be included in the analysis:

        a. Drought hazards and rainfall reliability?
        b. Periods of food shortages? If so, when?
        c. Seed shortages at planting time? When?
        d. Low prices for produce sold? Why?
        e. Limited possibilities for storing crops to fetch higher prices later?
        f. Possibility of labour shortages? When?
        g. Shortage of draught power during planting time?

    3. Ask them to consider which (group of) farmers would be most affected by the potential problems; in case information from the handouts is not sufficiently detailed to answer this, the groups should suggest additional information required.

- In the plenary session, one group presents the results of the first two assignments: their hypotheses with reference to the farmers' problems. The next groups add hypotheses that have not been mentioned earlier. Encourage discussion about the reliability and relevance of the data in the handout compared to asking the farmers' own views and opinions.

- A final discussion should address the third question. A list can be generated of additional information required to be able to differentiate between different groups of farmers and between men and women. Where would such information be available?

Variation:
- Relevant "real" documents or maps with information related to the participants' situation may be used instead of the Kenyan case. Different documents may be given to different subgroups for an analysis similar to the above.

Source: Walecka et al. (1987)

HANDOUT 2·1

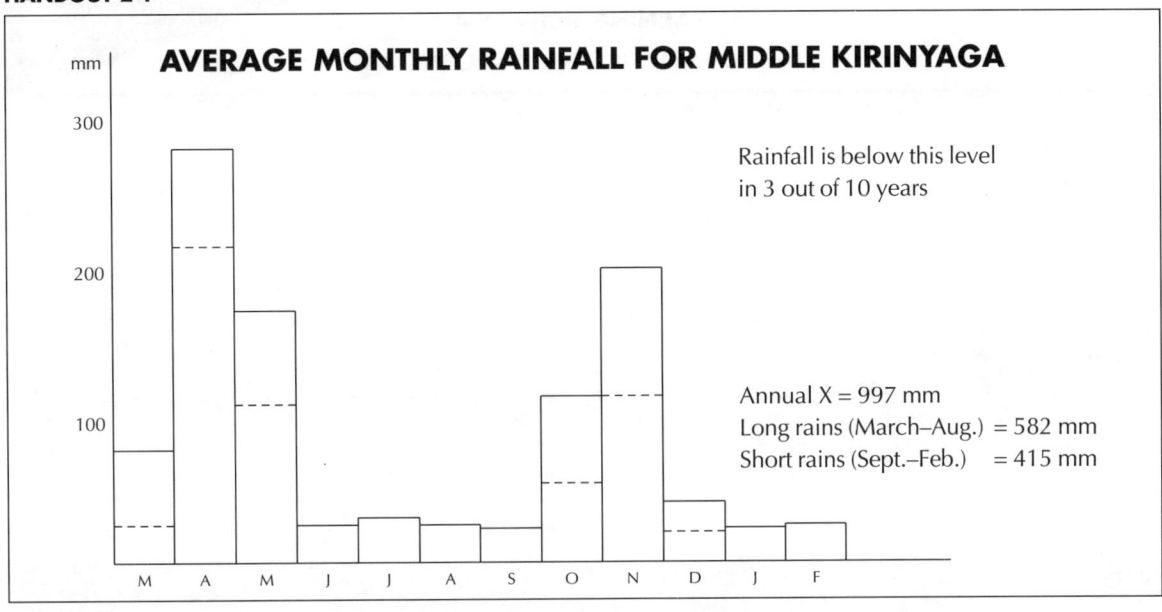

HANDOUT 2·2

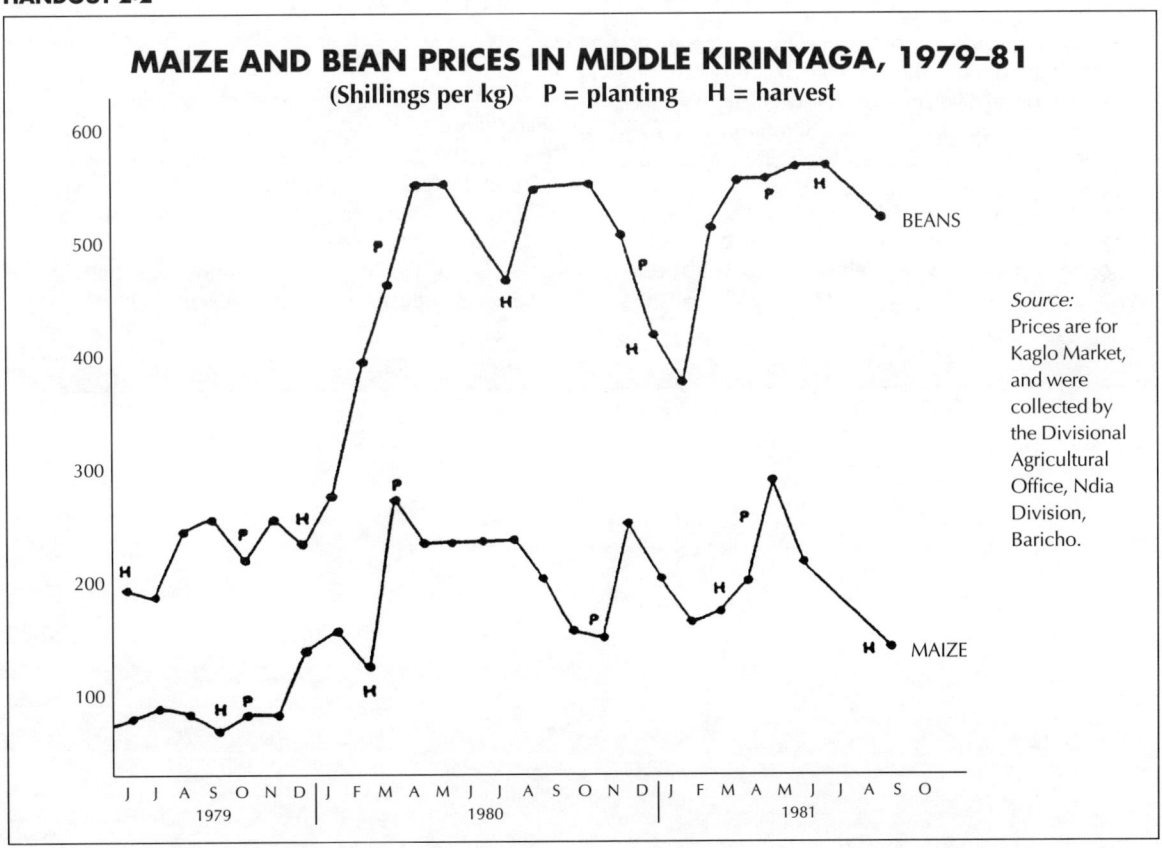

Source: Walecka et al. (1987)

LEARNING ACTIVITY 4
ROLE PLAYS OF FIRST VILLAGE CONTACTS

20–25 min per play

Objective:
- To understand the importance of open and motivating first meetings with key persons in a village and to develop the skills to hold such meetings.

Setting/approach:
- This activity gives the participants the opportunity to develop and act out simple role plays on first meetings between fieldworker and village key person.

Materials:
- None.

Procedure:
- Explain the objective of this activity and the general features of simple role plays: i.e. that there are only indications of different roles and no hard and fast rules, there is openness for individual interpretation and improvisation, and the main emphasis is on reflection afterwards on the scenario and results.
- Introduce the first situation to be played: a bad example of first contact between fieldworker and village key person.
- Ask the group to brainstorm briefly on the "don'ts" of such first contacts; or how should the fieldworker act in the example?
- Put forward a situation and divide roles as realistically as possible: e.g. a team of 2 fieldworkers visits a village chief to ask permission for activities.
- The volunteer actors may prepare themselves for a few minutes.
- The play should not take more than 10–15 minutes.
- Reflect briefly on the play, probably generating more don'ts; the final question is "what was the effect of the fieldworkers' bad behaviour?". You may want to focus the attention back to the importance of these first contacts.
- Follow the same procedure for a good example.

Variation:
- Instead of generating, in a brainstorming session, a list of don'ts and do's prior to playing the respective plays, the different players may be asked to create and show these in their own roles. The reflection period on the plays will then need more time to make explicit the mistakes or the lessons learnt in the play.

LEARNING ACTIVITY 5
EXPLAINING PTD

1–2 hrs

Objectives:
- To enhance participants' awareness of their own motivation for PTD, and of how to link this with farmers' interests.

- To strengthen participants' skills in interacting with farmers towards a joint understanding of PTD and overcome the "shopping-list" syndrome.

Setting/approach:
- Simulation in a workshop setting where a facilitator or participant plays a farmer asking for services rather than PTD.

Materials:
- None.

Procedure:
- Instruct one participant or a facilitator to play a farmer being approached for involvement in a PTD activity; this farmer knows that to deal with "development-outsiders" is usually to ask and insist on services; he may be interested in other ideas to improve his farm but does not know that this is an option and, above all, he gets angry when the outsiders try to evade his straightforward questions (e.g. for credit or inputs).

- The farmer takes the "hot seat" and one participant volunteers to sit with him and discuss the idea to start up PTD activities and his possible interest in joining these.

- Stop the discussion after arguments are being restated and discuss what happened, the approach taken by the outsider; other participants may take the seat one by one to try another approach.

- Ask the participants to summarise together the main lessons of this session and write these on the board or newsprint; a conclusion may emerge that certain farmers may never be interested in being involved with PTD; many others will, if their own concerns are not ignored and possibilities for new forms of activities that are complementary to them are introduced gradually.

Source: adapted from Scheuermeier & Sen (1994)

LEARNING ACTIVITY 6

SIMULATION: "VILLAGE UNKNOWN"

3–4 hrs

Objectives:
- To understand the importance of open and motivating first meetings with village key persons, and to develop the skills to hold such meetings.

- To be aware of important organisational and institutional issues when entering a new village.

Setting/approach:
- This simulation provides the participants with an opportunity to practise different steps in entering a new, unknown village. Its strength is its complexity, enabling many issues to become clear.

Materials:
- A general description of the situation for all participants. A set of directions for the different players.

- These directions should be developed on the basis of the participants' own reality. The best option would be for the facilitators to prepare the basic idea and set-up beforehand and finalise these together with the participants. An example of a Village Unknown simulation is given in Box 2.2 to stimulate facilitators' thinking.

Procedure:
- Explain the objectives of this learning activity in terms of agreed learning needs, or the importance of first contacts to ensure the effectiveness of the later PTD process.

- Explain the nature of a simulation play as a "close-to-reality" learning activity and clarify some of the basic rules: use general directions and descriptions for the play; each player is to improvise according to his/her own interpretation/experiences, but do not over-play!

- Distribute the general situation description of "Village Unknown"; ask participants to amend it where necessary to make it more realistic.

- Ask the participants to divide the roles: the PTD team (3–4 members), and the villagers (6 or more, include men and women). If there are more than 15 participants, it is advisable to do the simulation in two parallel groups, rather than have a large number of observing participants.

- Give the players 10–15 minutes to divide tasks and roles in each group and prepare themselves.

- The first visit of the PTD team should not take more than 30 to 40 minutes. After this the team may take 15 minutes to prepare carefully their second visit, which may take up to one hour if the team needs this much time.

- Provide for a well-organised reflection on this simulation, as a great number of learning points may have arisen during the game. Central questions may include:
  - Did the villagers, after the visit(s) of the team, understand what the PTD team wanted to do in their village? Who in the village did, who did not? Why did this happen?
  - Did the villagers really agree with the proposed programme? Who did, and who did not? Why?
  - Was there a feeling of mutual trust? Why or why not?

- Finally prepare together a list of main learning points from this exercise, e.g. in terms of what the team in a next case should or should not do.

Note to trainers:
In plenary sessions after simulation plays, the first possibility to speak should be given to the players who had the most difficult roles, i.e. the PTD team. They should be followed by the villagers, with the comments of possible observers given only at the end of the discussions.

BOX 2-2 EXAMPLE OF A SIMULATION PLAY: "VILLAGE UNKNOWN"

Information for all players:

Village Unknown is a poor farming village with 250 inhabitants. It is almost 3 hours' walk from the nearest town. The majority of the population are Christians and there is a small church served by the local priest.

In the town are the offices of a local NGO, which is affiliated to the church. The priest of Village Unknown met a staff member of this NGO once at a church meeting, visited the NGO offices, and was so impressed by its programme that he decided to send a letter to the NGO asking for its help to develop his church and the village. The NGO answers the priest that, on a particular day, a team will visit him to discuss his proposal and possible ways to start a programme in his village. On the agreed day, he awaits the arrival of the team in his house, together with a prominent member of the congregation. The other villagers are working in the field or at home.

Instructions for the PTD Team:

Your organisation adheres to a participatory approach in rural development similar to the one known as PTD. Based on the letter of the priest of Village Unknown, it is therefore decided to send a team to the village to establish first contacts and try to come to some kind of agreement/protocol on future cooperation. Of course, an important question-mark is the actual intention of the local priest. Is he hoping to get subsidies for his church, or is he really interested in the development of his village? And how does he relate to the other villagers?

As this village is wholly unknown to the team, it is thought that at least two visits will be required for them to reach a conclusion as to whether there is scope for a PTD development programme, and to arrive at a general strategy to start the programme.

Proceed now to draw up a plan for your first visit — what are you going to do? who do you want to visit? for which questions are you seeking answers? Develop also preliminary ideas for a possible follow-up visit. This visit will be planned in more detail on your return from the first visit to Village Unknown.

Instructions for the Villagers:

The villagers should play the following roles. For each role play, directions are given. However, the participants should further develop each role according to their own interpretation of the situation in Village Unknown.

- **The local priest:** You have requested the team to come. Generally you are interested in development of the village (not only your congregation), but you have very traditional ideas about development work: the NGO surely has a well established, ready-made programme? Your role will probably be to convince the farmers to do what the NGO wants, as you know them well.

- **Prominent member of the congregation:** You are one of the richest people of the village. In general you are more interested in upgrading the church than developing the village as a whole. The roof of the church, for example, urgently needs repair.

- **Village headman:** You are the official representative of the government. Unfortunately you are not aware of the present contact between the priest and the NGO. Generally, you suspect that the church has become a more influential body in the village than the government. Being a government representative, you are only familiar with a top-down approach in development work. Recently you have been informed that the MoA plans to start a maize improvement programme in Village Unknown.

- **Female primary school teacher:** Apart from being the teacher in the primary school, you are also the wife of the village headman and a member of the church council. Through your position you could have a lot of influence in the village. But, unfortunately, you are being dominated by your husband, the village headman — especially in public. All previous development programmes in Village Unknown have failed, because somehow the people quickly lost interest. You have, however, no idea of possible alternatives, but would probably welcome them very much if they were properly explained.

- **Other villagers, the farmers:** You are either working in the field or doing odd jobs at home. Of course you know that you were born in a poor village, working the land as you have learned from your fathers. And of course you would welcome help to do something, but previous development programmes have always failed, which has led to widespread frustration.

  Some of the villagers are Christians, others not. The first group look to the priest for help, the others look to the village headman. Unfortunately, experience has shown that these two groups often obstruct each other's initiatives. Anyway, what does it matter? Your opinion will never be asked.

The villagers will organise their village before the play starts: the local priest and his prominent church member in his house; the village headman also in his house, with his wife there too, working in the kitchen. The other villagers take places either in the field, working, or at home. Attributes may be used to add to their role.

UNIT 2.B / PARTICIPATORY SITUATION ANALYSIS: STARTING POINTS

OVERVIEW OF THIS UNIT

EXPECTED RESULTS

This unit aims to familiarise participants with the basic concepts and considerations of participatory situation analysis. These include the need for such analysis, its limitations and its general process, as well as the possible pitfalls and biases in implementing it.

After completing the learning activities in this unit, the participants are expected:

- to be aware of the need for participatory situation analysis and its limitations;
- to understand the general process of such analysis as well as its constraints and opportunities;
- to be sensitive to the differences between problems and solutions, problems and symptoms, and problems and their causes;
- to be aware of all possible biases when discussing problems with farmers and knowing ways to prevent these biases arising.

MAIN CONCEPTS

- **Participatory situation analysis:** Supporting (specific groups of) farmers to study and reflect on their local situation in order to identify constraints to sustainable agricultural development and opportunities to overcome these constraints.
- **Symptoms vs problems:** What is stated as a problem is often a symptom of an underlying problem which is only discovered and understood through further analysis.
- **Priority problem:** One that needs to be addressed first, because it affects many (of the targeted) farmers in an important agricultural activity, thus causing severe losses of income or production.
- **Outsiders' biases:** Tendency of outsiders to note only the situation of a selected few, e.g. master farmers, men, farmers living close to the main road, etc, and only selected issues.

TRAINING METHODOLOGY

A number of "classroom" learning activities are described here to introduce important conceptual issues related to situation analysis and problem diagnosis. Unit 2.C gives suggestions for learning the use of the relevant tools and methods. As the main focus of attention is on the differences in problem perception and analysis between outsiders with a "scientific" background and farmers, some kind of direct participation by farmers in the classroom sessions is strongly recommended.

LEARNING ACTIVITIES:

1. Recognising farmers' needs
2. The role of farmers in problem identification
3. Farmers as teachers
4. Problem or solution?
5. Problem–cause–problem: a case study
6. Problem-tree analysis

DISCUSSION

THE NEED FOR PARTICIPATORY SITUATION ANALYSIS

A participatory programme relates its activities directly to a felt need of the beneficiaries and takes their situation as a starting point (see Unit 1.C, pp. 41–8). A joint, participatory situation analysis is essential as it:

- helps outsiders as well as farmers better understand the local situation;
- generates ideas and options for future joint activities;
- strengthens the capacity of the village for critical reflection and analysis in current and future development;
- lays the foundation for subsequent farmers' control of and participation in future activities.

COMMUNITIES ARE NOT HOMOGENEOUS

Communities are not homogeneous and different interests exist alongside each other (see Units 1.E & 1.F, pp. 59–66 & 67–74). As most PTD programmes aim to work with the more marginal sections of society, this has important implications especially when analysing problems, needs and priorities at the village level. There must be:

- an **awareness** among staff of the differentiation in communities – differences in crops grown, different agro-ecological conditions, different economic groups, landowners vs tenants and the landless, pastoralists and sedentary farmers, men and women, young and old;
- a capability for **finding out** who is who in the area; specific tools exist, such as wealth ranking and social

TABLE 2·2
FARMER DIFFERENTIATION MATRIX

SEX	SOCIO-ECONOMIC LEVEL			AGE
	Stratum 1: poor adults	*Stratum 2:* intermediate status	*Stratum 3:* rich adults	Young people
Women	SG1	SG3	SG5	SG7
Men	SG2	SG4	SG6	SG8

SG = Social Group
Source: *Gianotten & Rijssenbeek (1994)*

diagrams (see next unit); easily observable local indicators may be found: quality of the house for economic status, number of cattle, etc;

- a willingness to **take those differences seriously**; first of all, in the situation analysis itself (whose problem?) but also in implementation; unfortunately, there are no hard rules here; many PTD programmes specifically target marginal groups but find themselves forced to pay considerable attention to the village élite, if only to protect the programme; others try to mediate between conflicting interests to find "win–win" situations.

In the latter approach it is important to check whether activities proposed by one group are not harmful to another. Gianotten & Rijssenbeek (1994) identified the relevant groups within communities in their programme area in Bolivia, facilitated a situation analysis with each group leading to proposals for activities, and systematically checked these with the other groups. A simple matrix (see Table 2.2) helped the team to keep track of this process.

PROBLEMS AND OPPORTUNITIES
Participatory situation analysis often focuses on problems felt by the (group of) farmers involved: production constraints, socio-economic factors and socio-cultural ones. The search for problems basically follows these few steps:

- brainstorming to provide a list of problems;
- looking for underlying causes via problem– cause– problem analysis; development of a "problem-tree" may be a useful tool to structure such analysis (MDF, 1990; see also Learning Activity 6, p.112);
- choosing a priority problem to act upon.

There is need for caution, though, in focusing on problem analysis:

- directly asking people their most pressing problems is often not very effective; it generates well-known "shopping lists". Good questions need to be found locally that lead to a critical discussion of local issues, e.g. questions asking for differences between present farming and that of 15 years ago and why one is less successful (Gubbels, 1988), or for the topics currently discussed most frequently by men and by women respectively;

- both farmers and PTD facilitators need to understand the difference between what is a *symptom* and what is an underlying *problem*. Poor yields may be a problem for the community whereas in fact this may be a symptom of the underlying problem of decreasing soil fertility, which in turn may be a symptom of ... etc. One should also be sensitive to the fact that what may be perceived as a problem by outsiders – for example, late planting by farmers, leading to water stress in the crop at the end of the season – may in fact be a *solution*, i.e. the farmer is planting late to cope with a labour shortage problem during the early planting season;

- many good activities may be found not by analysing problems but by being open to *ideas*; what ideas do people have, and why have they have not been able to implement them?

- a participatory analysis of one week, or even one month, will not provide outsiders with a total understanding of local key-bottlenecks to development; it will give only indications. For example, village socio-politics will take longer to understand. But the analysis may be sufficient to identify one, perhaps relatively small, activity to be tried out, providing both farmers and outsiders with an opportunity to further increase their understanding of what is required and possible. The concept of such "entry-point" activity is further discussed in Unit 2.D (pp. 129–34).

The criteria for choosing which issue should be given priority for action may include "technical" aspects as well as socio-cultural, political or strategic ones.

In the first category, the following are often used:

1. How many farmers are affected by a particular problem, i.e. are actually involved with the enterprise concerned and experience the problems, and are they targeted by the project?
2. How important is the affected enterprise to the farming/livelihood system?
3. How severe is the problem, and how great is the loss of production or income due to the problem?

Other farmers' criteria may include:

- are we capable of handling the problem more or less ourselves?
- are there traditionally specific groups that handle the issue, or are there any other local rules?
- is there political room for the (group of) farmers to handle it (marketing, landrights).

PREVENTING BIASES

In working in the community, it is crucial to avoid certain biases (Jiggins & de Zeeuw, 1992; Chambers *et al.*, 1989), including:

- **road bias:** confining exploration to those fields and households which are easy to reach;
- **élite bias:** restricting contact to the better-off farmers;
- **gender bias:** meeting mainly male farmers and not women farmers;
- **production bias:** concentrating only on production and neglecting post-harvest preservation, processing and food preparation.

These biases can be avoided by, for example, including women colleagues in the PTD team; inviting women to act as community guides; developing a rough map of the community together with guides, then visiting each quarter; visiting every household during community walks.

The danger of all these biases, of mixing problems with solutions, and of the differences in perception between farmers and outsiders that were discussed in Unit 1.G (pp. 75–83), all call for a systematic combination of various tools in the situation analysis. This enables continuous cross-checking to be carried out. As the answer obtained in one situation may be wrong or incomplete, the same question needs to be asked in another form, or of another person, or using another method. This is called triangulation.

FARMER-LED ANALYSIS

The most important challenge at this stage of the PTD process is to ensure that the analysis really is carried out by the (targeted group of) farmers themselves. There is often a temptation for outsiders to act as central "researchers" – to take home data generated, to indulge in interesting data analysis and reporting, and to draw conclusions. Reporting the results of such analysis back to the farmers is only partly a solution.

Giving farmers the central role requires sensitivity to their real involvement and control at all times. This may mean leaving original data in the village, asking farmers for permission to publish them, or giving farmers a credit in the publication.

LEARNING ACTIVITY 1
RECOGNISING FARMERS' NEEDS

1½–2 hrs

Objectives:
- To increase awareness of the difference between what participants, as outsiders, think that villagers need and what are, in fact, the villagers' real felt needs.
- To enhance understanding that facilitators have as much to learn from villagers as villagers have to learn from them.

Setting:
- Any large, quiet meeting place.

Materials:
- Paper and pencils (enough for all participants).

Procedure:
- Ask the group to discuss the following questions:
  - to what extent do you think villagers are aware of their problems?
  - if they are not aware of some problems, then are those problems real?
  - what does it mean to say that someone has a problem but isn't aware of it?
  - who determines when a problem is a *real* problem?

- Set up a role play situation (based on a situation known to most participants) between two individuals or two subgroups: one representing villagers, the other representing urbanised development workers. After a short preparation, the development workers explain to the villagers what they think the villagers' main problems are and how they might be solved. The villagers respond to these ideas and suggestions, offering their own analysis of their problems.

The process is then reversed: the villagers explain what they think are the main problems in urban areas and how they might be solved. The development workers respond.

- Ask group members who are not involved in the role playing to be observers. Divide these observers into two groups: one to record the views of the villagers, the other to record the views of the development workers.

- After the role playing is finished, ask the entire group to discuss what they have learned about themselves and their attitudes towards villagers.

- Each observer-group reads its list of ideas; those expressed by the "villagers", and those expressed by the "development workers". Ask them to discuss what they have learned about *real* villagers from this activity. What would they have to do to really understand villagers' attitudes?

Source: Crone & St John Hunter (1980)

LEARNING ACTIVITY 2
THE ROLE OF FARMERS IN PROBLEM IDENTIFICATION

1¼ hrs

Objective:
- To increase awareness of the importance of the role of farmers in situation analysis and problem identification.

Setting/approach:
- This is an organised confrontational discussion divided between some participants playing the role of farmers and others playing the role of outsiders.

Materials:
- Pen and paper for participants; facility to visualise the main points of the plenary discussion (e.g. blackboard and chalk).

Procedure:
- Explain the objective of the activity in terms of clarification of the respective roles of farmers and outsiders in situation analysis.
- Ask the participants to form an even number of small groups to prepare themselves for the plenary discussion. Half the groups should prepare themselves to "defend" the position of farmers, the other half to "defend" the position of PTD facilitators. Central questions are:
  - why is the role of farmers (or PTD facilitators) essential in situation analysis and identifying problems?
  - what is the specific role of farmers (or PTD facilitators) in the analysis?
- After 10–15 minutes, moderate a plenary discussion starting with the second question; conclude this discussion by asking whose role is the most important – farmers' or outsiders' – and why.
- The discussion may be concluded by summarising the main points on the blackboard. The central role of farmers versus the facilitating, catalyst role of the PTD facilitator is likely to be a main theme.

Variations:
- This activity can be made more complex, but more realistic, by asking one farmer group to focus on men farmers and another group to focus on women, and/or by suggesting that one PTD group focus on extension staff and another on researchers.
- This learning activity can be moderated by the participants themselves, providing an opportunity to practise their skills in this respect. For example, two participants may be asked to prepare themselves for the plenary discussion while the small groups are working. One of them may concentrate on moderating the discussion while the other collects and later presents the main points made.

LEARNING ACTIVITY 3
FARMERS AS TEACHERS

2–3 hrs

Objectives:
- To increase awareness and acceptance of the importance of farmers' knowledge and insight, and to sensitise the participants on the subjectivity in assessing these.
- To enhance the participants' sensitivity to the differentiation among farmers and its impact on situation analysis.

Setting/approach:
- In a workshop, participants interview invited farmers in small groups.

Materials:
- None, other than pen and paper for participants.
- The farmers (4–6 depending on the number of small groups) play an essential role in this activity. They should be well briefed on the objectives of this activity and their role in it. They should all be involved in the agricultural activity that is to be discussed during this session. The second objective can be reached only if farmers with different backgrounds are invited, e.g. men and women farmers.

Procedure:
- Introduce the participating farmers and explain the general procedure of the activity and the role of the farmers in it.
- Ask one of the farmers to describe in the plenary session one important agricultural enterprise in his or her area (e.g. the growing of maize) as well as the difficulties and problems presently encountered in this enterprise (about 20 minutes).
- The participants form small groups to prepare themselves for interviewing a farmer, with the following instructions:

1. List all the problems mentioned in the presentation which seem to affect the discussed enterprise.
2. Prepare questions to analyse the problems raised with one of the farmers in order to find the underlying causes and their interrelationships.
3. Be prepared for the fact that, by the end of the interview, one priority problem has to be selected to be taken up for further action.

- After 15 minutes, ask the small groups to interview one farmer each; this may be an informal discussion (30 minutes) but should conclude with the group choosing one particular problem as being the priority problem for further action.

- In the following plenary session, ask each group to report its choice of priority problem and to explain the reason for that choice. The subsequent discussion may focus on the following questions:
  - the reasons for differences in the choice of priority problems (subjectivity of interviewers, their specialism, incomplete data leading to different interpretations, misunderstanding, incorrect interview techniques, etc);
  - the different background of the farmers: to what extent has this led to a different choice of priority issue?
  - the incompleteness of a method identifying priority problems through information from one, subjective, source only. What biases played a role?

Variations:
- If only one or two farmers are available, the interview part may be replaced by asking the questions in a plenary session.
- Extension officers of the area may play the role of the farmers, but that is often less effective than the above.

LEARNING ACTIVITY 4
PROBLEM OR SOLUTION?

1 hr

Objective:
- To understand the difference between statements by farmers that present real problems and farmers' statements that refer to solutions, or symptoms.

Setting/approach:
- This is a relatively simple exercise, in which a handout with a list of statements is studied to identify which statements present real problems.

Materials:
- Paper and pencils, blackboard and chalk, or newsprint and markers, and the handout "Is this a problem?" (p.109).

Procedure:
- After explaining the aims of this exercise, ask the participants to form small discussion groups.

- Distribute the handout, or another one prepared especially for the workshop based on local conditions, and ask the participants to read it.

- The groups then proceed to answer the questions in the handout and list their answers, with reasons, on newsprint.

- After 15–20 minutes the groups return to the plenary session. One group is asked to present its results extensively whereas the results of the other groups are used to review, or comment on, those of the first group. Where different responses occur, a good discussion will be essential to clarify the difference between a problem and a solution. A concluding question will be: how to handle this in the field?

Variation:
- If time is short, the participants can be led through this exercise in the plenary session. This will take about 20 minutes.

Note to trainers:
Consider the following assessment of the 8 statements in Handout 2.3:

1. Describes a farmer practice, not a problem.

2. Possible problem, seemingly inefficient use of land and capital since the large amounts of applied fertiliser are not significantly increasing yields.

3. A problem, limiting production.

4. Not necessarily a problem as there is no indication that weeds are limiting production seriously; farmers may have other priorities for their capital than buying tools, if they are available.

5. A possible problem, limiting production.

6. Not a problem: a possible solution to a possible income problem.

7. Inefficient use of land or labour, under-exploitation of resources?

8. Not a problem: farmers planting late is their solution to the problem of low availability of tractors.

HANDOUT 2-3

IS THIS A PROBLEM?

Consider the points below then identify which of the following eight statements by farmers qualify as problems.

For each statement that you identify as a problem, note whether it is a problem because:

a. it limits production, or
b. it is an inefficient use of land, labour or capital.

1. "We are broadcasting our wheat, rather than using seeders."

2. "I apply over 200 kg of nitrogen with average yields of only two tonnes per season."

3. "Nearly half the groundnuts harvested have failed to germinate."

4. "We do not have proper tools to do a good job of weeding."

5. "Zinc deficiency limits the yields of maize here."

6. "We may increase our incomes if maize is intercropped with some beans."

7. "The major rains are followed by a minor rainy season, but I plant very little at that time."

8. "We always plant late because of low availability of tractors."

Source: adapted from Walecka et al. (1987)

LEARNING ACTIVITY 5
PROBLEM–CAUSE–PROBLEM: A CASE STUDY

2 hrs in small groups, 45 min in plenary session

Objective:
- To increase awareness of the need to "dig deeper" into problems to find their underlying causes and their inter-relationships.

Setting/approach:
- This activity presents a simplified case study.

Materials:
- Paper and pencil; newsheet and markers, or cards with markers; a handout describing a case study plus instructions, e.g. the Zambia case study in Handout 2.4 (p.111).

Procedure:
- After explaining the objectives of this activity, ask the participants to form discussion groups of 3–4 people. Give a handout to each group. This may contain the Zambia case study, or a similar one based on local conditions.
- Ask the discussion groups to answer the following questions:
  - identify at least 5 problems in the case study and list them;
  - for each problem, list its causes as stated in the text;
  - describe the relationship between each problem and its causes, possibly in a diagram.
- Suggest that each group present its result in the plenary session using either the newsprint or the small cards. The second option would permit further manipulation during the plenary discussion to visualise the relationships between problems and causes.
- The final plenary discussion should focus on deepening understanding of the causal relationship between different problems. The implication is the need for "digging deeper", and continued questioning of the reason behind what is posed as the problem.

Note to trainers:
Some possible answers to the questions are given in Box 2.3. Note that the answers focus on technical issues typical of conventional agriculture. Other root causes may be brought into the discussion. Lack of disaggregation of data according to gender or economic status may also be noted and introduced for discussion.

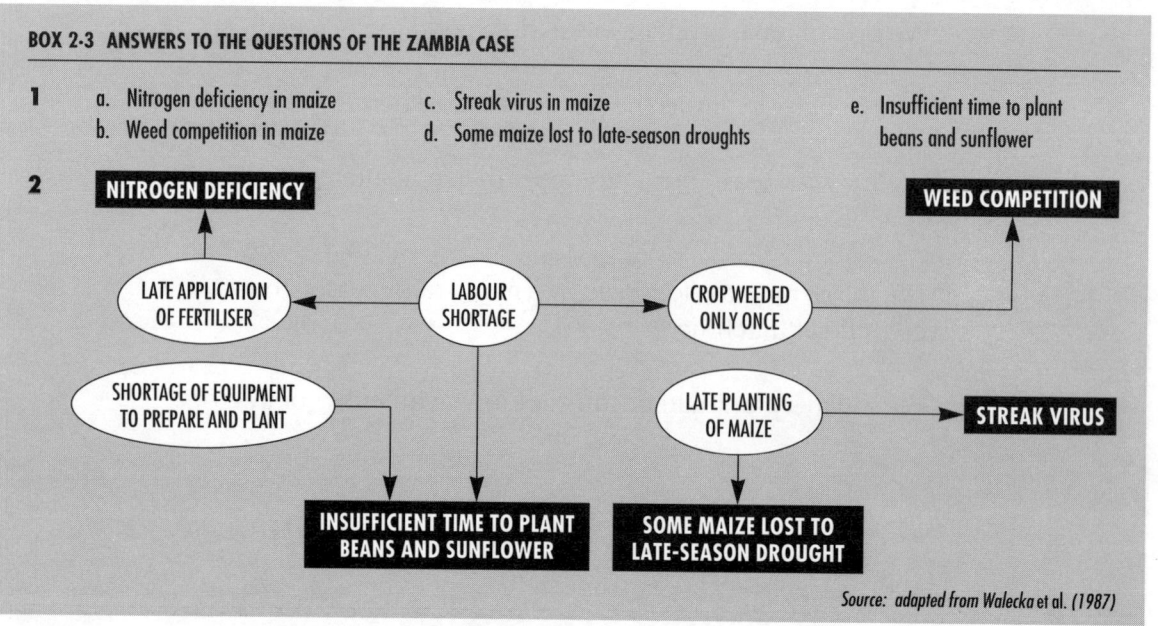

BOX 2-3 ANSWERS TO THE QUESTIONS OF THE ZAMBIA CASE

1.
   a. Nitrogen deficiency in maize
   b. Weed competition in maize
   c. Streak virus in maize
   d. Some maize lost to late-season droughts
   e. Insufficient time to plant beans and sunflower

Source: adapted from Walecka et al. (1987)

HANDOUT 2-4

MAIZE PRODUCTION IN KABWE DISTRICT, ZAMBIA

- Maize is the principal crop, occupying about 75% of the cultivated area.

- There is a good market for maize and adequate access roads to those markets.

- Farmers plant relatively large areas of maize, generally 5–10 ha.

- There is enough land to enable farmers to use a field for several years and then leave it fallow while they shift to another site.

- Planting takes place between October and December.

- Farmers must wait for the first rains before planting. They use ox-ploughs and plant following the plough. The planting may continue for two months or more because of shortage of labour and draft animals.

- About half the farmers have their own oxen, while the others have to borrow or rent draft animals.

- Fertiliser is applied to the maize, but it is often applied late because of labour shortages at planting time. This affects the efficiency of the fertiliser. There is evidence of nitrogen deficiency in the crop.

- Weeding is begun after planting has been completed. The crop is usually weeded only once and there is evidence of weed competition.

- The principal insect in maize is stem borer. Farmers apply insecticide when the problem is particularly serious.

- The principal disease problem is streak virus, which is transmitted by leafhoppers and affects late-planted maize.

- Rainfall is uncertain during the cropping season; there is particular risk of late-season drought.

- Other crops in addition to maize, such as sunflower and beans, are usually planted after the maize planting is completed. Farmers often do not have enough time to plant as much of these other crops as they would wish.

Source: Walecka et al. (1987)

LEARNING ACTIVITY 6
PROBLEM-TREE ANALYSIS

1–2 hrs

Objectives:
- To increase awareness of the need for "digging deeper" into problems to find underlying causes and their inter-relationships.
- To enhance skills in using the problem-tree technique for systematising discussions on felt problems.

Setting/approach:
- A workshop setting in which the participants practise problem-tree analysis.

Materials:
- Cards with problems in a case study on a bus company (Box 2.4).

Procedure:
- Introduce the problem tree as a tool for more systematic discussion and analysis of problems; mix the cards with problems of the bus company and hang them on the board; moderate a plenary discussion asking the participants to help order the cards in such a way that those with problems influencing other cards are linked to each other.

- Jointly reflect on the result: Are the cause–effect relations clear? How was this result obtained? What questions were asked? Does the position of a problem in the tree indicate its importance, for example, in terms of the problem to be addressed first?

- Ask the participants to form small groups focusing around a particular farming activity of interest; ask each group to:
  - brainstorm on the most important constraints faced by farmers in this activity and prepare a list of these;
  - write each constraint on a card and develop a problem tree; some well-focused support may help the groups to develop their skills in this;
  - display the results somewhere for all to see during breaks, etc.

- Ask the participants in a plenary discussion to reflect on the usefulness and the limitations of such problem analysis – in systematising their own analysis of local situations and/or supporting farmers in such analysis.

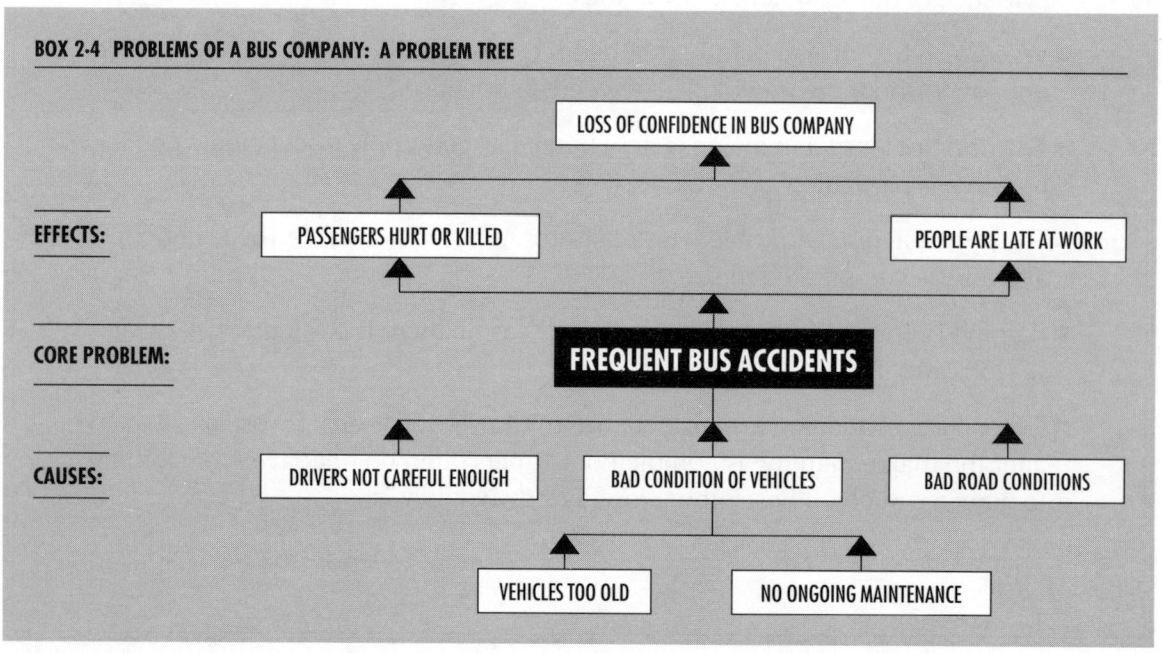

BOX 2-4 PROBLEMS OF A BUS COMPANY: A PROBLEM TREE

UNIT 2.C / METHODS IN PARTICIPATORY SITUATION ANALYSIS

OVERVIEW OF THIS UNIT

EXPECTED RESULTS

The last 10 years have brought an increasing number of approaches, methods, tools and techniques for analysing, together with farmers, their farming system, its potentials and constraints. The best way to carry out such a situation analysis would, in almost every situation, include a combination of many different methods and techniques. Central here are approaches that give farmers the leading role and that support their own analysis of existing systems. This unit makes the participants aware of the different methodology options presently available to carry out a situation analysis together with farmers. It aims at stimulating them to develop their own methods appropriate to their situation.

After completing the learning activities of this unit, the participants are expected:

- to be aware of the common features of methods of situation analysis that involve maximum farmer participation;
- to be familiar with the methods most often used in participatory situation analysis and to be able to apply selected methods with sensitivity to social/gender issues;
- to have increased creativity in designing methods and approaches for situation analysis with maximum farmer participation.

MAIN CONCEPTS

- **Participatory analysis:** Highly interactive methods and tools for working with farmers to increase joint understanding of the local situation, problems and opportunities.

- **Visualisation:** Information being creatively displayed in the open, for all to see, work with, manage and control.

- **Land literacy:** Methods and tools to support farmers and others in observing or "reading" the land and the environment to monitor sustainability.

- **Important methods, tools and techniques** for participatory situation analysis.

TRAINING METHODOLOGY

Studying the issues of this unit should preferably start with reflection by the participants on the methods of situation analysis used so far in their projects, and the problems encountered. This could be followed by a discussion of criteria to identify methods of situation analysis that aim to maximise farmers' participation. A handout based on the discussion section of this unit may help to give participants an overview of existing approaches and methods.

The development of skills for using the methods and techniques mentioned is a continuous process. Any training strategy should give opportunities for the alternation of practice and guided reflection. Although the best learning situation is in the field, there is ample evidence that prior skill training and workshop tryouts are important in preparing participants for their fieldwork.

Workshop tryouts are a simulation in which participants practise the methods and techniques in situations that reflect real life; for example, through role plays of different interview situations or exercises in mapping and ranking. The challenge is to create a situation which is as realistic as possible. In certain cases, farmers may be invited to the workshop, but a stronger learning situation can be created if the exercise focuses on real issues/problems felt by the participants themselves during the workshop.

The recent PLA Training Guide (Pretty *et al.*, 1995) contains a further collection of training ideas and exercises appropriate in stimulating learning on methods for participatory situation analysis.

LEARNING ACTIVITIES:

1. Methods of participatory situation analysis
2. Good and bad practice
3. Role-playing interviews
4. Extensionists play farmers
5. Practising problem census
6. Practising techniques of participatory situation analysis (transects, maps, ranking)
7. Creative ways of posing problems
8. Visualisation

DISCUSSION

CENTRAL FEATURES

Until the mid-1980s most analyses focusing on rural situations in developing countries displayed some of the following characteristics (McCracken *et al.*, 1988): the duration of the analysis was long – sometimes several years – and the structure often fixed and formal; the scope was often limited and usually concerned with a single issue, ignoring wider interlinkages and implications; the integration was weak, even in multi-disciplinary teams; the depth of investigations was quite exhaustive; the direction was essentially top-down, i.e. dealing mainly with government agencies and institutions, and only indirectly with the farmers and landless; the level of participation of the local farmers, and even local researchers and decision-makers, was low; and the cost was considerable, with low efficiency in time and manpower.

Such methods of analysis are inappropriate within the context of a PTD development programme. In such a programme, the methodology used to enhance understanding of the local situation should aim at maximising farmers' participation and increasing their awareness and self-confidence in improving their own situation. In general, therefore, the methods chosen are:

- "simple", to be controllable by farmers;
- "quick", to prevent frustration and loss of interest;
- aimed at knowing only what is really needed;
- sensitive to social and gender differentiation;
- "informal" – taking place at farmers' fields or houses, to make them accessible;
- made up of group sessions alongside interactions with individual farmers.

Participatory Rural Appraisal (PRA – for further reference see Theis & Grady, 1991; Chambers, 1992, and Pretty *et al.*, 1995) employs a wide range of methods to enable such participatory analysis. In Africa, the GRAAP (GRAAP, 1987) and DELTA (Hope *et al.*, 1984) approaches have gained regional importance. The latter includes an important emphasis on enhancing farmers' understanding of socio-economic developments influencing the sustainability of their livelihood.

The participatory analysis has three equally important purposes:

1. enhanced understanding by the farmers and outsiders of the local situation, its main problems and potential for the development of sustainable forms of agriculture;

2. increased awareness of the farmers, and the establishment of a relationship of trust and understanding between farmers and outsiders as a basis for joint activities;

3. strengthened capacity of farmers and communities for critical analysis of local developments and problems and opportunities.

Most methods include some form of visualisation: issues raised are not just spoken about or noted down but are "put up" for all to see, by means of words on a large sheet of paper, symbols, drawings, pictures, etc (Box 2.5).

BOX 2·5   THE ROLE OF VISUALISATION

Visualisation of information and issues raised:

- gives memory support
- helps to enhance understanding
- increases accessibility of information for all present
- enables joint ownership and control
- provokes critical reflection and awareness
- makes learning enjoyable
- enables the sharing and spreading of results

It may not be effective if:

- too many details are included
- there is insensitivity to the dangers of misunderstanding
- the approach decided upon does not take into account the local situation, the aim of the exercise or the skills of staff involved

---

A multitude of methods, tools and techniques are now available for analysis of the local situation with farmers. In the following overview, they are grouped into methods for interactions with farmers, and tools and techniques that can be used during such interactions.

INTERACTION METHODS

Participation in village life
PTD practitioners actively participate in the activities undertaken by the farmers. These may be agricultural operations or household work, but also cultural or religious events. Especially at the early stages of programme development, this is a very appropriate approach which is often overlooked.

Direct observation
This includes any approach which relies on directly observing objects, events, processes, relationships or people in the field, and keeping a record mentally and/or in note or diagrammatic form. This serves mostly as a basis for later discussions with farmers and/or resource persons. Observation may take the form of actual measurement, through which "invisible" processes sometimes become visible. Counting of the number of certain pests and their predators in rice, for example, helps farmers and outsiders to understand better how they spread and could endanger the crop. Such observation and "reading" of the land and the environment, also known as "land literacy" (Campbell, 1994), is crucial in the search for sustainable forms of land use.

People's observation is often strongly coloured by their background, education and discipline. It is a challenge for PTD practitioners to go beyond this and keep an eye open for a variety of important features. Some of the more innovative forms of direct observation rely on carefully chosen indicators. These are events, objects, etc that are easily observed or measured and can be used as an indicator of some other variable that is more difficult to observe; for example, house type as an indicator of wealth (McCracken et al., 1988).

Semi-structured interviewing
This is a form of guided interviewing where only some of the questions are predetermined and new questions or lines of questioning arise during the interview, in response to answers from those interviewed.

Semi-structured interviews may take various forms:

- with individual farmers to obtain representative information;
- with key informants to obtain special knowledge;
- with groups to obtain information on community interactions.

These interviews may be "chance encounters", i.e. take place whenever and wherever a group or individual is met and there is an opportunity to chat, for example during a walk through the village. Alternatively, they may be carefully arranged, with those interviewed selected according to certain criteria (sex, age, poverty, occupation, etc) or at random (Conway et al., 1987).

The crucial factor in making an interview a success, rather than simply demonstrating the interview skills of the PTD practitioner, is the basic attitude of the interviewer towards the farmers and their culture and way of life (see Unit 1.E, pp. 59–66). The PTD practitioner will also need communication skills to enable open dialogue to take place (see Unit 1.G, pp. 75–83).

Although, in general, no formalised questionnaire will be used, the PTD team will very carefully prepare the main issues to be discussed as a kind of checklist, and learn this list by heart. Preparation of this list is based on information already obtained about critical issues in the farmers' situation. Particular attention may be needed to find relevant questions linking day-to-day experiences of farmers to longer-term sustainability of the farming system. Further guidelines for semi-structured interviews, be they individual or group interviews, are given in Box 2.6 (p. 116).

Structured interviews and surveys
In structured interviews, almost all questions are predetermined, often in the form of a questionnaire, and are generally asked one by one by the interviewer. They may be used to collect detailed information on certain issues – for example, details of household economics, or occurrence of different pests in maize. They are generally less effective in the first stage of participatory situation analysis but may be needed at later stages. Farmers should be involved in the decision on the use of this method and its subject, and can also be involved in designing the questionnaire and holding the interviews.

Group problem diagnosis
Here we refer to all participatory analysis methods that have the following features in common:

- interaction with groups of farmers with a common interest;

BOX 2-6 USERS' NOTES ON SEMI-STRUCTURED INTERVIEWING

DOs

- **DO** spend time preparing a comprehensive interview guide or checklist. Write it down... for guidance during interviews.
- **DO** remember that the interview is structured by the team for a purpose.
- **DO** be relaxed but serious.
- **DO** explain clearly who you are.
- **DO** let each team member finish their line of questioning.
- **DO** probe a topic by using the 6 helpers: what, when, where, who, why, how.

  Also use the key probes:
  - how do you mean?
  - tell me more about that
  - anything else?
  - but why?

- **DO** also probe by asking informants to role-play: "Suppose..."
- **DO** judge the responses – are they fact, opinion or rumour? Ask yourself: What qualifies the informant to give me that response? Also evaluate the reliability of the interview.
- **DO** take a neutral attitude, listen carefully and pay great attention to non-verbal behaviour.
- **DO** record the interview, by taking notes in detail either during or afterwards.
- **DO** pay attention to the selection of informants. Use participatory maps or wealth rankings to ensure a good mix of informants.
- **DO** record the names of the informants.
- **DO** be open-minded: be prepared for bad and good interviews. If it is going badly, conclude politely and leave.
- **DO** pay attention to team dynamics by holding regular meetings and brainstorming sessions. These are often as important (even more so, sometimes) as the interviews themselves.

DON'Ts

- **DON'T** interrupt each other.
- **DON'T** accept the first answer – probe all topics.
- **DON'T** ask leading questions. Many questions that can be answered with a "yes" or "no" are leading questions.
- **DON'T** interrupt informants.
- **DON'T** supply answers for an informant who is hesitating.
- **DON'T** dominate the proceedings by using inappropriate non-verbal behaviour.
- **DON'T** take up too much time of an informant who is busy.
- **DON'T** show disapproval or distate about local conditions or about drinks or food offered.
- **DON'T** indicate disbelief by criticising, or even just smiling.
- **DON'T** ask questions that combine two queries – e.g. "Do you have a medical centre here and are you happy with it?"
- **DON'T** ever let the informant feel cross-examined.
- **DON'T** ask about sensitive information in front of a group of onlookers.

Source: adapted from Pretty (1990)

- facilitation of a group process, taking place over one or more sessions, following different steps in problem analysis;
- aiming at raising farmers' awareness and motivation through the group process;
- providing a basis for possible collective action.

To enable intensive interaction and reflection, the discussion groups generally have only 5–10 members.

Crouch (1991) has elaborated the Problem Census, in which a facilitator asks farmers to form small groups of 4–6 members, to identify what they perceive as main problems in their farming. Each group reports its results to the others. The facilitator then assists in a plenary session to cluster similar problems and arrive at one common list. The farmers return to their small groups to identify priority problems and report these back to the main group. This leads to a common list of the most important problems. The meeting ends with a decision on further action, e.g. how to focus the next meeting on possible solutions or on further analysing the root causes of the problems identified.

A group problem discussion can be made much more focused by using problem-posing materials in the form of posters, slides, drama, etc. The preparation and use of such materials is discussed below.

TOOLS AND TECHNIQUES

In all the above methods, different tools and techniques may be used to facilitate the discussion, to promote the participation of everyone involved, and to enhance understanding of the issues at hand. Details on using these can be found in recent PRA publications, e.g. Theis & Grady (1991) and Pretty *et al.* (1995).

Diagramming: construction of conceptual models

A conceptual, or diagrammatic, model is a simple schematic device which presents information in a readily understandable form (Conway *et al.*, 1987). The power of such a model lies in the fact that it visualises the information being discussed so that all can see it and improve or elaborate on it, and presents it in the form of a model showing relationships that may previously have been unclear.

Diagrams are most effectively used when produced together with, or by, the farmer(s) and subsequently discussed. Different groups of farmers (in terms of sex, age, socio-economic status) may need to be involved, as they often emphasise different aspects. Sometimes, diagrams are prepared by the PTD team and then presented to the community for discussion.

Diagrams of space include maps, for example of a village, a watershed or a farm. These maps usually present only selected information relevant to the issues under discussion. Social maps, for example, may show different characteristics of the village population, such as family or clan distribution, land-tenure patterns, and socio-economic status; they can be an important tool in increasing insight into the differences within a village.

The preparation of simplified farm maps, or farm models, is a powerful tool in analysing constraints and potentials in the farming system. If movement of biomass within the farm is indicated, such models become bioresource flow diagrams. The number of species in the diagram is an indicator for species diversity and the number of movements of biomass an indicator for internal recycling (see Box 2.7).

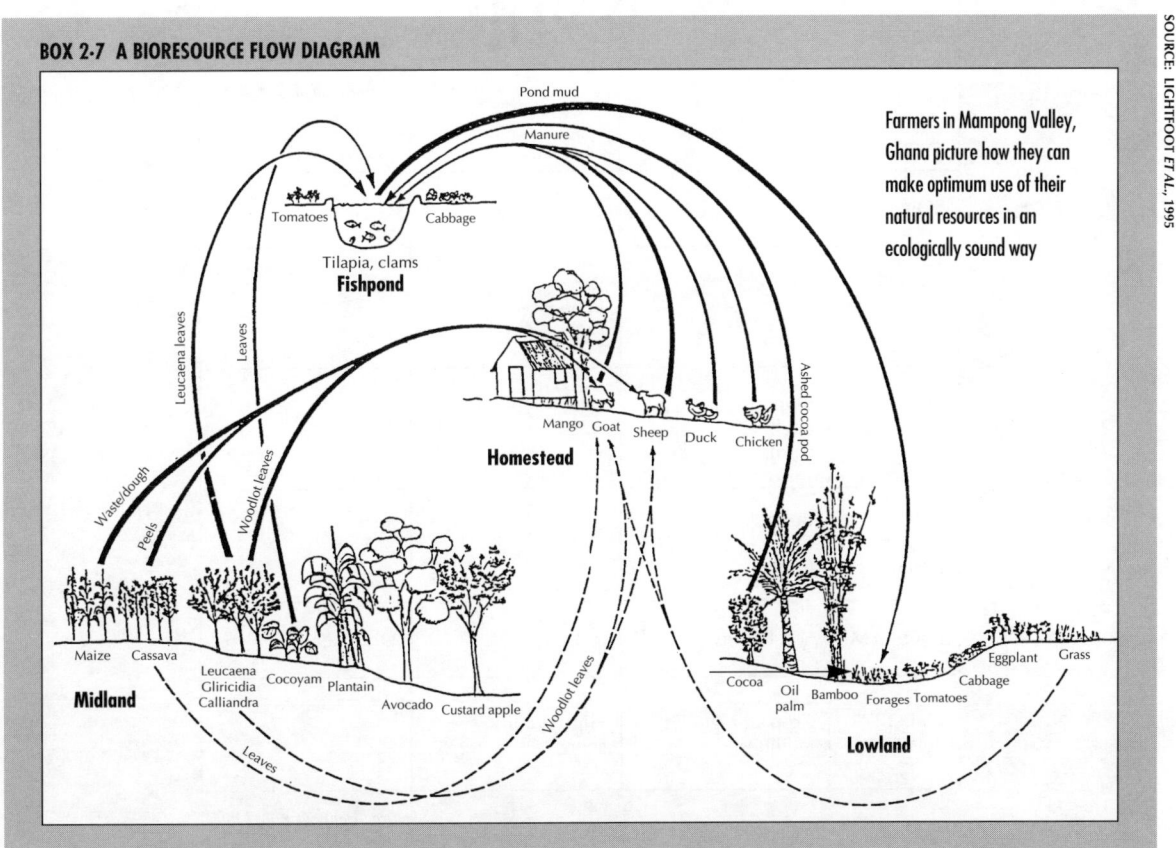

BOX 2-7 A BIORESOURCE FLOW DIAGRAM

Farmers in Mampong Valley, Ghana picture how they can make optimum use of their natural resources in an ecologically sound way

SOURCE: LIGHTFOOT ET AL., 1995

Transects are a simple technique used to identify the different land-use systems in the area and provoke discussion on specific problems for each of them. The team walks to the periphery, exploring differences in land use, vegetation, soils, cultural practices, infrastructure, trees, water availability and so on. The transect diagram produced is a stylised representation of a single or several walks (see Box 2.8).

Diagrams of time include seasonal calendars, showing, for example, sequences of agricultural practices and other activities throughout the year. Differentiation for various groups of farmers is often necessary (see Box 2.9).

Sequences of events over a greater number of years are often analysed in so-called time trend diagrams. Other conceptual diagrams include, for example, cause–problem diagrams such as "problem trees" and "objectives trees" (see Unit 2.B, pp. 102–12).

Construction of physical models

The preparation of real-life physical models is often more time-consuming but, at the same time, it provides farmers and outsiders with concrete information as a basis for joint analysis and discussion (Hahn, 1991). These are especially relevant for the analysis of interactions between farms and the wider environment and may even cover one complete watershed.

BOX 2-8 A TRANSECT

LAND TYPE	UPLAND	MIDLAND	"CHAUR" LOWLAND
SOIL		Silty loam	
CROPS		Rice Wheat Mungbean	
TREES	Mango Litchi Citrus		
LIVE-STOCK & FISH	Bullocks Goats Cattle		Fish
PROBLEM	Soil fertility		Drainage
OPPOR-TUNITIES	Crop storage facilities	High quality seeds	

Source: Lightfoot et al. (1990)

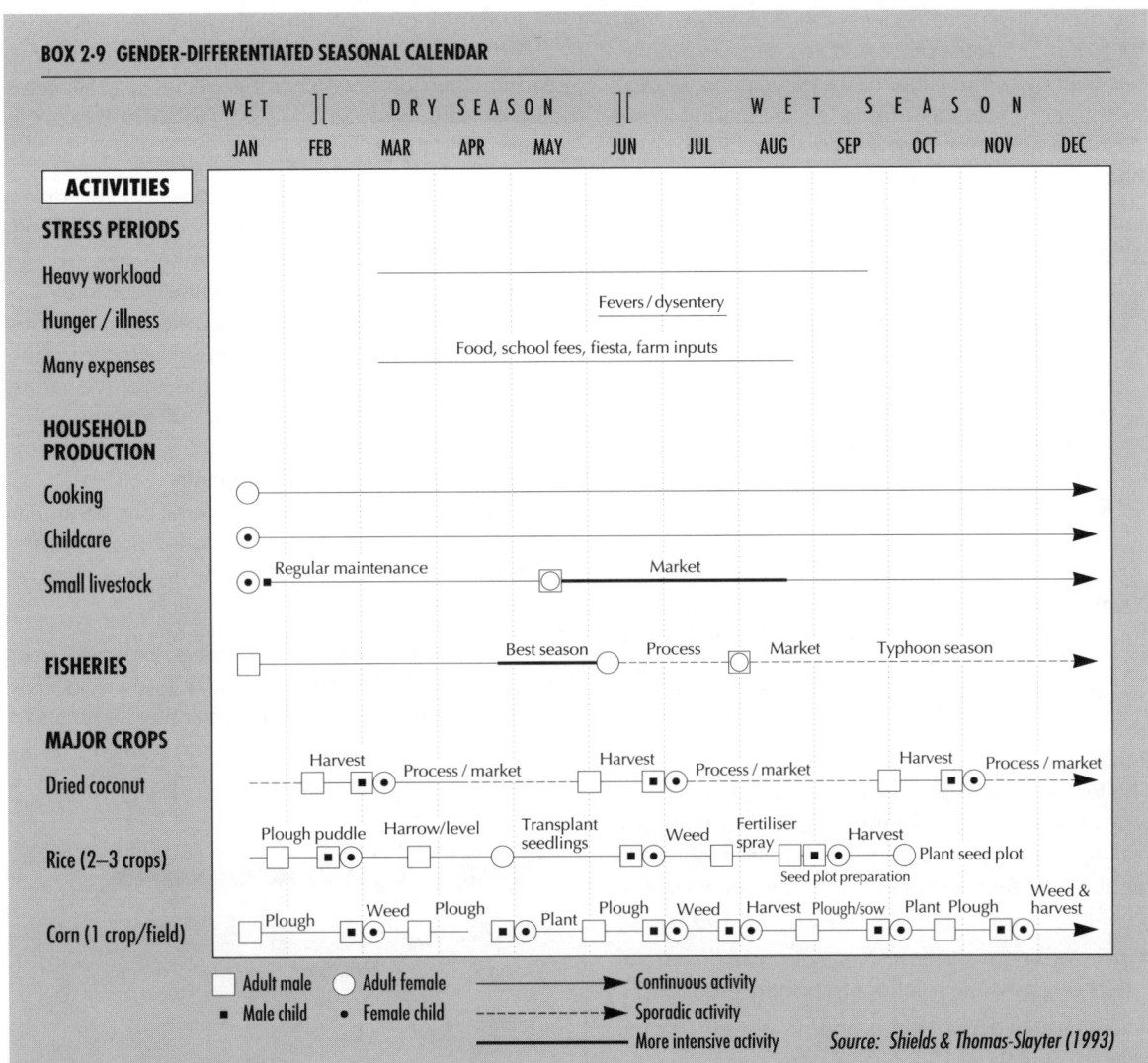

Problem-posing materials

Problem-posing materials are visual aids that help participants to reflect on the problem under discussion and express their feelings about it. Examples of such materials are posters, flannel boards (GRAAP, 1987), slides (Engel, 1991), short stories, drama (Adoyo, 1994), songs, etc. They are effective only if they show a concrete example of the problem being discussed, in such a way that it would be familiar to most participants, and provocative. The aim is to raise questions, not to provide solutions. Problem-posing materials, or "codes", are very different from other visual aids, such as illustrations, which merely help the teacher to explain a certain issue.

To be effective in stimulating discussion and raising awareness, a code should (Hope *et al.*, 1984):

a. deal with a problem about which the farmers have strong feelings;
b. show a very familiar scene in everyday life;
c. use contrasts or action to raise awareness and questions;
d. focus on one problem so that the discussion goes deeply only into this particular problem;
e. be simple, clear and visible;
f. avoid distracting details;
g. stimulate the interest and touch the hearts of the group.

The strength of using codes is that it initially invites the group to discuss other people's situations rather than their own, thus allowing the participants to speak more freely on certain sensitive issues. It also focuses the discussion immediately on real situations, rather than hypothetical scenarios.

A group is confronted with the problem-posing material and guided through a discussion that generally follows five stages:
1. *Description:* What do you see in it? What do you think people are doing, and feeling?
2. *First analysis:* Why are people doing what they are doing? (The group still talks about the poster, etc.)
3. *Real life:* Does this happen in our real life? Can you give examples?
4. *Related problems:* Are there other problems related to the "symptom" of the code? (But sometimes one should stick to the first one, if that is really the central problem.)
5. *Root causes:* The group is then challenged to analyse the situation at a much deeper level. The main question often repeated will be: "But why?"

Ranking

This technique can be done in at least two ways: pairwise ranking and matrix ranking. Both aim at quickly making farmers' priorities and preferences explicit and highlighting their criteria for such preferences. The technique is useful (Jiggins & de Zeeuw, 1992) for:

- inventorising available resources and identifying and selecting possible solutions to priority problems;
- helping outsiders to understand farmers' criteria and decision-making;
- pinpointing desired characteristics to be taken into account when selecting materials for field trials;
- establishing criteria for evaluating the results of experiments.

In *pairwise ranking*, farmers are asked to compare, for example, species of trees two at a time and indicate why they prefer one species above the other. This will result in a list of criteria that determine the preference of the farmers present, as well as a preference list of the species discussed.

In *matrix ranking* (to use the same analogy), a series of relevant species is listed on the ground, or across the board, while a series of criteria to evaluate them (generated, perhaps, through pairwise ranking) is listed down the side (Box 2.10). In a group discussion, farmers then judge which of the species (or other items listed) is the best, second best, etc, judged by each criterion in turn (Theis & Grady, 1991).

An important tool in analysing socio-economic differentiation in communities is known as wealth ranking. Names of all the households are written on cards and selected villagers are asked to sort the cards individually into as many piles as there are wealth categories according to them. Great attention is paid to the wealth criteria for each pile. Because of the sensitive nature of the subject, this tool must be used carefully.

Case histories and critical incidents

Farmers are asked to describe in detail one particular event, incident or history as a starting point for discussion on a particular problem, or an innovation by farmers themselves.

A case history may focus on farmers' experiences in trying out and adapting a particular innovation. This is effective in revealing trends over time, sources of new ideas, and records of tried and proven experiments.

BOX 2·10 AN EXAMPLE OF DIRECT MATRIX RANKING

Criteria	TREE SPECIES			
	Eucalyptus	Palm	Acacia	Pine
Fuelwood	4	1	2	3
Building	4	1	2	3
Fruit	1	4	2	3
Medicine	4	1	3	2
Fodder	3	–	4	2
Shade	4	3	1	2
Charcoal	2	–	3	4
TOTAL SCORE	22	10	17	19
RANK	A	D	C	B

If you could plant only one species, which tree would you choose?
EUCALYPTUS

4 = best   1 = worst

Repeat for a number of interviewees.

Source: Theis & Grady (1991)

Critical incidents focus on a key event in the past to study a subject of particular concern (e.g. pest control in rice). Farmers trace the events that led up to it, the nature of the incident, responses to it and its consequences. Group discussion of this event may lead to further understanding of this problem and identification of ideas for testing (Flanagan, 1954).

SELECTING AND COMBINING METHODS, TOOLS AND TECHNIQUES

Any participatory situation analysis will always combine several of the methods and techniques mentioned above, if only to cross-check information generated from other sources. Such cross-checking is essential when qualitative (and sometimes intuitive) appraisal methods are used. The challenge is always to be creative and open for possibilities to increase farmers' participation in and control of the analysis. When selecting an appropriate mix of methods for a particular situation, one should consider the following:

1. Is the main focus of the situation analysis on the result, the data collected, or the process of trust-building and awareness-raising? In the first case, there may be a need for more measurement and structured interviewing; in the latter, more actual participation and group dynamics may be required.

2. Should the data generated mainly serve to increase the understanding of the PTD team, or should their main purpose be to enhance the collective understanding of the farmers?

3. What skills do the staff carrying out the situation analysis have? Leading a problem-census group process or preparing effective problem-posing materials is likely to be more difficult than using a questionnaire in a structured interview.

4. What potential is there within the community? To what extent can people read and write, use conceptual models, speak up in small or larger meetings? Remember to differentiate between men and women, leaders and the poor, elders and the young.

LEARNING ACTIVITY 1
METHODS OF PARTICIPATORY SITUATION ANALYSIS

1¼ hrs

Objective:
- To enhance understanding of the central features of the methods used in participatory situation analysis and to enable maximum farmer participation.

Setting/approach:
- A brainstorming session for participants to reflect on the methodologies used so far in situation analysis in the course of their work.

Materials:
- Materials for visualising the results of the small groups in plenary session: a flipchart with markers, or blackboard and chalk.

Procedure:
- After explaining the purpose of this activity, ask the participants to discuss, in pairs, for 5 minutes: "What methods and techniques have you used in your work to find out the actual situation of farmers and their problems?".

- Ask the groups to report on one method or technique each; hold second and third rounds for groups to report any methods that have not been mentioned. Suggest that one of the participants write the main points on the blackboard for all to see.

- Have the participants form small discussion groups (of 4–6) to share their experiences of problems encountered in using some of these methods.

- After approx. 30 minutes, ask the groups to report their results, writing the main points on the flipchart or blackboard; an opportunity should be given for brief open discussion to clarify problems reported by the groups.

- Ask the participants to reformulate the problems mentioned in positive language – as requirements/criteria for selecting or designing better methods of situation analysis.

Variation:
- The groups can assess the effectiveness of the methods used in terms of generating relevant information as well as enabling farmers' participation and control over the process. What are the characteristics of the more effective methods?

LEARNING ACTIVITY 2
GOOD AND BAD PRACTICE

1¼ –1½ hrs

Objective:
- To increase awareness of the difference between good and bad interviewing techniques.

Setting/approach:
- Participants are confronted with a number of photographs or other images showing interview situations and together generate guidelines for good interviewing.

Materials:
- A series of 6–12 photographs or slides, each showing different interview situations demonstrating good and bad practices. These images can be locally made or taken from reports. Blackboard and chalk to visualise the results of the plenary session.

Procedure:
- After explaining the general procedure of this activity, ask the participants to form small groups of 4–6 people.
- Circulate the numbered photographs among the small groups and ask them to study each one carefully and note the good and bad points in each situation.
- The small groups come together in a plenary session and the participants formulate guidelines for carrying out interviews on the basis of what they have learned by studying the photographs.
- Collect these guidelines on the board in the form of Dos and Don'ts.

Variation:
- The good and bad examples of interviewing can be dramatised by the facilitators or, better still, by small groups of participants. Participants will need time to develop the "plot" and prepare themselves to be able to show clearly the main issues of the play (compare Learning Activity 3, p.123).

Source: Pretty et al. (1995)

LEARNING ACTIVITY 3
ROLE-PLAYING INTERVIEWS

1 –1½ hrs per "play"

Objectives:
- To increase awareness of good and bad interviewing techniques.
- To enhance participants' skills in conducting successful interviews.

Setting/approach:
- Participants practise interviewing. A learning situation or "play" is created, which is as realistic as possible. The details of each play and its setting are developed spontaneously with the participants.

Materials:
- Auxiliary materials for the play; pen and paper for observers (or observation forms in the variation).

Procedure:
- Explain that the idea of this activity is to simulate an interview situation as realistically as possible (individual or group, probably semi-structured).
- Together with the group, develop a *realistic* "play" setting in which an issue is discussed during the interview that is convincing for the person being interviewed: for example, a participant who is known to have problems in promoting participatory approaches in his or her project may be interviewed by other participants in order to identify the main constraints s/he faces.
- Ask 2–4 participants to form the interview team, and give them 10 minutes to prepare themselves. This preparation may include:
  - main questions to be asked/points of interest, as a checklist;
  - division of tasks among the team members.
- The team proceeds with the interview while other participants watch and are asked to make a note of good and bad practices (30 minutes).
- Moderate a plenary discussion on how the interview proceeded, focusing on what the team did or did not do. First ask the team members to share their experiences; the interviewee may then comment on these, and finally the other participants.
- Collect the main points and visualise them on the blackboard. A list of dos and don'ts may be prepared if this has not been done before.

Variation:
- Participants may be asked to pose as farmers, but this may only be taken seriously if they have their own farm and do not have to fabricate one for sake of the play.

LEARNING ACTIVITY 4
EXTENSIONISTS PLAY FARMERS

2–3 hrs

Objectives:
- To increase understanding of how and how not to conduct an interview.
- To enhance skills for carrying out "good" interviews.
- To promote awareness of the role of extension workers in problem analysis.

Setting/approach:
- This is a simulation exercise where extensionists, or other individuals with extensive knowledge of the local situation, pose as farmers in analysing local problems.

Materials/other requirements:
- For a group of 24 participants, six extension agents posing as farmers; possibly observation forms.

Procedure:
- Explain the objectives of this activity with special reference to the need to integrate the issues and skills studied in the previous activities of this unit.

- Ask the participants to form small teams to carry out the interviews, one for each of the resource persons posing as farmers; note that the interview will consist of two "rounds".

- Suggest that the following steps be taken:

  a. Teams meet to discuss their interviewing strategy. They discuss who will introduce the group to the "farmer", what will be said to the "farmer" as an introduction, what the rules will be as to which member asks questions, etc. Using the interview guidelines they drew up in the previous exercise, they decide on a plan to address these issues. Advise them that their first interview should be fairly general, to get an overview of the farming system. In the second interview, they should focus on selected topics and problems of interest (10 minutes);

  b. The first interview can take place anywhere – outdoors, in separate rooms, etc. Participants should pretend that the interview is taking place at the farmer's house (20 minutes);

  c. In the small interview teams, the participants review both the form and the content of the interview. They plan topics to discuss during the second interview, which will take place with the same "farmer". Thus they gain practice in focusing on problem areas identified but not well understood during the first interview. Meanwhile, the "farmers" meet to discuss and list the strengths and weaknesses in the interviewing techniques of the trainees (20–30 minutes).

  d. In the second interview with the same "farmer", the team uses the information from the first interview to explore topics in greater depth (20 minutes);

- Each team discusses its interview, selects a reporter, prepares a short summary of the content of the interview and criticises its own interviewing technique. The "farmers" also meet in a group, select a reporter, and prepare a critique of the interviewing methods used on them. They evaluate the technique by stating how comfortable they would have felt during the interview if they had really been farmers (15 minutes).

- Each team gives a 10-minute report on the content and technique of their interview (50 minutes);

- A representative of the "farmers" presents their report (10 minutes).

Variation:
- Inviting farmers instead of extension workers to be interviewed has the advantage of more direct communication, i.e. there is no need for "acting". It may also help to further open the eyes of the participants to the wealth of knowledge farmers have. And by asking the farmers to comment on the way the participants conducted their interviews, at the end of the activity, the farmers become teachers of the participant – an important role reversal.

Source: based on Walecka et al. (1987)

LEARNING ACTIVITY 5
PRACTISING PROBLEM CENSUS

2 hrs

Objective:
- To increase participants' skills in moderating a problem-census group process.

Setting/approach:
- The participants are given an opportunity to moderate a group discussion in an "as-real-as-possible" situation. They should be familiar with the issues discussed in Unit 1.G, pp. 75–83.

Materials:
- Pen and paper; observation forms.

Procedure:
- After explaining the purpose of this activity, ask the participants to form discussion groups of 8–10 persons. Give each group the following instructions:
  - choose among your group two moderators for the discussion to come;
  - give the two moderators 10 minutes to prepare themselves;
  - discuss a commonly felt critical issue, following the stages of the problem-census method: identify different problems/constraints, analyse their root causes, choose a priority and try to develop possible solutions (about 1 hour).
- Back in the plenary session, the groups report on their discussions, with special reference to:
  a. *the results:* have the root causes been found? have good solutions been identified?
  b. *the process:* has everybody been involved? how was the influence of the moderators? what was good and what was bad practice?
- The selection of the issue to be discussed needs special attention. More can be learned if a topic can be chosen which is an actual "hot" issue within the group of participants. For example, topics concerning the content of the workshop where there is still great disagreement among the participants ("the need to work via the élite to reach the poor", "the need for special women-oriented programmes", "cooperation with NGOs or the government", "constraints in research-extension linkages", etc), or problems related to the workshop process itself (low level of participation of many of the participants, unclear relevance of the course to the actual work situation, poor facilities in lodging and food).

Variations:
- During the workshop, there are often other opportunities for the participants to practise moderating plenary discussions. If these are used for skill development, care should be taken to provide time to evaluate the role of the participant-moderator.
- Some participants may act as observers in each group. In the final plenary discussion, they will then be able to present more detailed feedback on the role of the moderators.

LEARNING ACTIVITY 6
PRACTISING TECHNIQUES OF PARTICIPATORY SITUATION ANALYSIS (TRANSECTS, MAPS, RANKING)

1–2 hrs

Objective:
- To increase skills in preparing and using different techniques of participatory diagnosis.

Setting/approach:
- The participants are given an opportunity to practise using different techniques within the "safe" workshop environment.

Materials:
- Generally pen and paper, large sheets for drawings, blackboard and chalk, handouts with guidelines on the use of the chosen technique.

Procedure:
- Explain and discuss the chosen technique in a plenary session; e.g. through a short lecture, or by distributing a handout prepared from the discussion section of this unit and/or the literature indicated, and discuss it.

- Creatively organise a tryout of the chosen technique; here are some ideas:
  - ask the participants to form small groups and make transects of the complex around the workshop centre or of the workshop building;
  - form groups of 10 participants and ask one in each group to moderate a matrix ranking exercise; e.g. to find preference among all the techniques of diagnosis so far discussed (across the board) on the basis of criteria (to be identified, along the board); or on simpler topics such as preference for types of 4-wheel drive vehicles, places to go on holidays, etc;
  - ask them to do a wealth ranking with a colleague participant, analysing, for example, the wealth status of all the staff in his or her office (30–50 people).

- Hold a plenary session to:
  - discuss the good and bad points of the different trials to find where improvements are needed;
  - assess together the potential and limitations of the technique used.

LEARNING ACTIVITY 7
CREATIVE WAYS OF POSING PROBLEMS

2–3 hrs

Objective:
- To increase participants' skill in creatively using a variety of techniques to prepare problem-posing materials.

Setting/approach:
- Participants are asked to prepare problem-posing materials in small groups and use them in the larger group, in a workshop setting.

Materials:
- Tape recorders, photographs or drawings of people in rural communities, newsprint and coloured markers, anything else to create the problem-posing materials.

Procedure:
- Divide the participants into 4 groups.
- Ask each group to discuss a real problem situation of farmers from the experiences of one of the group members (20 minutes).
- Ask each group to spend 30 minutes on planning how to present the problem situation discussed to the larger group. Each group must use a different "medium", for example: drama, story, poem, song, cartoons, posters.
- After each group has presented its code, the viewers identify the issue/problem as they see it; the presenting group then describe the problem as they see it.
- Moderate a final plenary discussion focusing on, for example:
  - how could the techniques used be improved next time? did they have the potential to provoke a heated discussion?
  - what are the relative advantages of one technique as compared to the others? when to use which?

Variations (not mentioned in the source):
- There are many other possible ways to make problem-posing materials (slides, puppet show, etc). The advantage of those presented here is that they can be prepared relatively quickly using materials at hand.
- This activity will gain even more if the participants are asked to prepare codes for real "problems" actually experienced within the group as it attends the workshop (group interaction, facilities, time pressure) or problems participants generally experience in their work (bureaucratic procedures, lack of influence in decision-making).
- When participants' real problems are the focus of this activity, the presentation of each code should be followed by a discussion, using the steps of description, first analysis, real life, related problems and root causes, as described on p. 120.

Source: Ellis (1983)

LEARNING ACTIVITY 8
VISUALISATION

2 hrs

Objectives:
- To increase awareness of the importance of using visualisation in participatory situation analysis.
- To understand the pros and cons of using visualisation in working with farmers.

Setting/approach:
- In an interactive workshop session, participants are stimulated to think critically about the use of visualisation.

Materials:
- Blackboard with chalk or newsprint and marker; audiovisuals showing visualisation in practice, such as the video by ICLARM on Pictorial Modelling – for details, see Appendix, p. 222: ICLARM (1991).

Procedure:
- Ask participants, either individually or in pairs, to express why visualisation could be important in participatory situation analysis.
- Collect their answers on the board and challenge participants to find the main themes; compare with Box 2.5, p.114.
- Together, watch the ICLARM video or another similar audiovisual.
- Moderate a plenary discussion, possibly preceded by brief whispered discussions in simultaneous "buzzing groups" of 2–3 participants, on the following questions:
  - is the visualisation approach presented applicable in your situation? if not, why not?
  - what, in general, are the shortfalls of using visualisation (danger of misconception, poor handling, need for skills, acceptability at farm level, socio-culturally specific)? how to overcome these?

UNIT 2.D / LOOKING FOR THINGS TO TRY

OVERVIEW OF THIS UNIT

EXPECTED RESULTS

It is a crucial challenge for a PTD programme to maintain the participatory mode of working once joint analysis with farmers has made the priority constraints clear. The question "What to do now?" can be answered by supporting farmers (both men and women) in identifying and selecting, from various sources, ideas for innovations as "possible solutions" to be tried out.

After completing the learning activities in this unit, the participants are expected:

- to realise the complementarity between farmers' and outsiders' knowledge and their ideas in looking for concrete solutions;
- to be familiar with a variety of methods to determine options for experimentation and their own role in this process;
- to be aware of the importance of farmers' criteria in choosing between technical options and to know ways of eliciting these;
- to be able to reflect critically on one's own (organisation's) criteria and policies and the importance in choosing between the different options for action;
- to be able to facilitate decision-making on the agenda for experimentation, leading to a clear understanding of what is to be tried out and why.

MAIN CONCEPTS

- **Options rather than fixed packages:** Farmers prefer a range of possibilities to work with and adapt to their local situation rather than one fixed innovation, or a fixed combination of interrelated measures.
- **Farmer-based solutions:** Many ideas for improvements exist among farmers that can be included in selecting options to be tried out.
- **Eliciting farmers' criteria:** Discussion process in which the preferences and criteria of farmers are made explicit.
- **Entry-point activities:** Activities strategically chosen for implementation in the first year(s) of PTD programmes to ensure effective participation at the initial stages.
- **Hypothesis for experimentation:** Formulation of the idea to be tried out in a systematic, analytical way.

TRAINING METHODOLOGY

A full appreciation of the richness of farmers' knowledge and experiences in finding solutions for local problems can be reached only by direct interaction between participants and farmers. Fieldwork to inventorise innovations by farmers themselves should be a central element in learning PTD. This may be part of the workshop, or organised as a follow-up to be discussed in further training.

In "classroom" discussions, an overview of possible methods and techniques may be developed and the participants may practise some of them. A critical reflection on their own role at this stage of the PTD process as compared with conventional extension and research approaches will most likely touch, once again, upon the fundamental premises of the PTD approach.

LEARNING ACTIVITIES

1. Studying farmers' recent innovations *(compare also Unit 3.A, Learning Activity 2)*
2. Looking for solutions
3. Moderation of a creative brainstorming session
4. Eliciting criteria
5. Own priorities
6. Making a choice
7. Hypothesis formulation

DISCUSSION

MAINTAINING A PARTICIPATORY MODE

It is one of the main challenges for any participatory development programme to maintain the participatory mode of working after analysing with farmers the local situation. Frequently, at this stage, the experts give farmers the solution that has been identified to be implemented without delay. To meet this challenge, PTD emphasises that agricultural innovations are generally not ready-made solutions but should be treated in a specific situation as an object of "research": of experiments in which farmers play the central role. They are possible solutions for problems faced by the farmers – nothing more and nothing less.

Moreover, there is rarely *one* solution; for example, a fixed package consisting of different interrelated measures, appropriate for all farmers. Farmers with a complex farming system may be better served by an increase in the number of options they have, which they can use and adapt in various combinations under different circumstances.

GENERATING FARMERS' OPTIONS

Solutions to address the problems raised are often found among farmers themselves. This, in itself, is an important reversal in thinking in agricultural extension (compare discussion on indigenous knowledge in Unit 1.E, p. 60). However, such an idea may not have been put into practice for various reasons; for example, it was not known to a particular group, it was never linked to the problem under discussion, or it could not be implemented by an individual farmer without support from others or from an outside agent.

Throughout the process of interaction between farmers and resource persons, ideas on possible improvements will be mentioned, or can be solicited. The PTD facilitator should keep an open mind for such ideas and opportunities as they arise. The generation of possible solutions may also be the logical next step in a problem-census discussion with farmers (Unit 2.C, pp. 119–20). Groups may try to list ideas for possible solutions from among their own members as well as from the PTD facilitator. The main virtue of this approach is that the group develops its own strength and becomes the "carrier" of subsequent activities.

In other cases, or complementary to this, specific methods can be used in finding things to try. Jiggins & de Zeeuw (1992) have listed a number of them, including:

- **farmers' expert workshop:** a meeting of farmers acknowledged in the community to have knowledge and experience relevant to the problem under discussion;

- **innovators' workshop:** in a similar meeting, those farmers may be invited who are known to be generally innovative in running their farm. They may have ideas – which might even have been tried out on a very small scale – that are worth trying by the others;

- **study tours:** representatives of the farmers' group or of the community visit places/sources of possible innovations; for example, agricultural research stations, another development project, specific farmers known for their expertise;

- **exchange visits:** similarly, farmers from different villages involved in a PTD programme may visit each other on a regular basis. This is an important method of farmer-to-farmer extension and is discussed in detail in Unit 4.A: Cross-visits, pp. 196–7.

Case histories and critical incidents are among the techniques discussed in Unit 2.C, pp. 120–21, which may be used here as well.

Many sources of ideas are open to farmers

LINKING WITH IDEAS ELSEWHERE

The PTD facilitator's crucial role in all the above methods is to engender an atmosphere of creativity, enabling brainstorming with the participation of all involved, in order to activate the enormous resources of farmers' experiences.

Yet, the PTDer's role as a liaison with outside sources of knowledge is also important. PTD facilitators may bring ideas obtained during their education and training, from reading or from visits to research stations. Where possible, they stimulate direct contact between farmers and different sources of information, including research stations and other government services. Some programmes see this as their main role to ensure that problem-solving by farmers can proceed without continuous facilitation by outside agents (CARE, 1994).

FARMERS' AND OUTSIDERS' CRITERIA

To answer the question as to which of the ideas for solutions are worth trying out, a first step would be to make explicit the relevant criteria. Both farmers and outsiders or PTD facilitators have their own preferences and priorities. Only by trying to make these explicit will possibilities be created to evaluate the various ideas properly.

Farmers will use various criteria in different combinations. Unit 3.D (pp.177–8) presents a detailed discussion of criteria used to evaluate the results of the trials. Of course, farmers may have different priorities depending on their socio-economic position, or sex, or age (Box 2.11), and their preferences may change over time – for example, due to a change in household situation or in market conditions.

A PTD team has to exert a concerted effort to make farmers' criteria explicit. The basic approach is to confront the farmers with one or more options and ask them which they prefer and why (ranking). Their answers are then formulated in the form of criteria, and finally analysed jointly to find the farmers' reasons for choosing those criteria.

Ashby (1990) describes three ways to elicit farmers' criteria:

- **absolute evaluation:** each alternative option is judged on its own merits (e.g. what do you like/dislike in this variety? or: is this technique acceptable to you? why yes/why no?);

BOX 2-11 WOMEN FARMERS CHOOSE BEAN SEEDS FOR TRIALS

In the Farmer Participation in Technology Assessment Project (CIAT & IFDC) in Colombia, scientists asked selected innovative farmers to rank beans before planning trials. One apparently unattractive small grain type was chosen when women took part in the discussion. The wives recognised the bean as a high-yielding, flavourful variety that had disappeared from the locality; one that had proved to be profitable, and swelled on cooking.

Farmers who preferred large-grained beans were entrepreneurial and either unmarried or recently married and living in the extended family. Those who preferred small-grained varieties headed households with children. If women farmers had not been involved in the planning at an early stage, the variety and evaluation criteria preferred by them would have been omitted.

Source: Ashby et al. (1990)

- **pairwise comparison:** each alternative is compared successively with all other alternatives, or with the actual practice or local variety: the "baseline". If you had to choose between these two alternatives, which one would you select, and why? This can become tedious if applied to more than 4–6 options;

- **ranking amongst alternatives:** alternative technologies are ranked from most-preferred to least-preferred, meanwhile explaining why that ranking order is being chosen: why is each one preferred over the next in rank?

This process will eventually result in a list of possible options, on the one hand, and a clear list of criteria used by farmers to choose among these, on the other.

But the PTD team also needs to be explicit about its own (organisation's) criteria, to enable farmers to include these (if they wish) in their decision-making. The following questions need to be asked:

- What is the organisation's capacity to assist farmers in the different possible trials? Farmers may like to try out a new breed of cattle for which credit is expected, but this may not be available. Or technical support may be expected which the programme may not able to provide or to mobilise.

- Does the proposed activity contribute towards greater sustainability of the farming system? Does it lead to LEISA rather than HEIA development?

- What is the organisation's strategy, especially regarding suitable activities in the first year(s). It is often advisable at this stage to avoid taking up complex and risky activities; entry-point activities may be chosen which, for example (Vel *et al.*, 1989):
  - attract the attention of the poor and respond to their interests rather than those of the rich;
  - do not require many inputs and are relatively simple;
  - bring results in the short term, rather than in the long run;
  - provide a first step for farmers' organisation; and
  - give good possibilities for subsequent follow-up activities.

The PTD team may have to point to associated support structures required to make a certain option a success: transport, credit, and marketing facilities.

MAKING A CHOICE

In the final decision-making, the remaining serious options can be assessed with the criteria developed together using the matrix-ranking technique (Unit 2.C, p. 120). This would probably take place in some kind of farmers' group meeting. In practice, however, the decision-making process may be less democratic, depending on local customs and tradition. Communities often have their own mechanism for taking decisions, in which the village chief or informal leaders may play a decisive role. While being aware of this, and therefore paying special attention to ensuring support from those leaders, it is a basic challenge for the PTD team to find ways to include, from the beginning, less vocal groups, such as poorer farmers or women, in the decision-making process.

A dilemma arises when the idea preferred by the farmers does not make sense in the eyes of the PTD facilitators, extension workers or researchers. After reasons on both sides have been discussed openly, it is PTD's fundamental orientation that the final choice lies with the farmers. Investing in supporting experiments that go against one's own views will strengthen farmers' sense of ownership and provide important learning opportunities for them (Schmitz *et al.*, forthcoming). Ultimately this may prove to be more effective than trying to impose one's own views.

HYPOTHESES FOR RESEARCH

Once one particular option has been chosen to be tried out during the next season, a final step prior to the actual trial would be to formulate this option in terms of a hypothesis. This is a crucial step in the dialogue between farmers and outsiders. It helps the farmers to define more exactly what they want to try out and why, and enables them to analyse more clearly the results of the trial. It is a planning, monitoring and evaluation tool. It helps both parties to understand each other's logic better. It provides an opportunity to check the reasons for the problem and prevents jumping to conclusions about possible solutions. Lack of clarity at this stage will make people stumble in formulating a hypothesis.

Such hypotheses may have a very simple format. The term "hypothesis" may not even be used in the discussions, thus focusing on the basic logic alone.

BOX 2-12 HYPOTHESES FOR A FARMER-LED TRIAL ON ZERO GRAZING

IF: More farmers keep their animals in zero-grazing units

THEN:
1. Diseases will be better controlled
2. Crop yields will be higher
3. Conflicts with neighbours will decrease
4. Milk yields will be higher
5. Children will have time to go to school
6. Hooves will have to be trimmed
7. Carrying forage will make more work

BECAUSE:
1. We will see the animals more often
2. We will be able to use more manure
3. There will be no crop damage by grazing animals
4. It will be easier to feed supplements
5. Herding will no longer be needed
6. Animals will not be able to walk down their hooves
7. Animals will not be able to collect forage themselves

Source: adapted from Pretty (1990)

A simple format would be:
- if…   I plant in rows rather than at random
- then…   I will get higher yields
  - …   etc
  - …   etc
- because… I will able to control weeds much better

Box 2.12 gives an example of such hypotheses for a trial on zero grazing.

Ideas chosen for testing may also be summarised in "idea sheets" (see Box 2.13). Leaving these with the interested farmers gives them a simple tool to check whether what has been decided is being done.

BOX 2·13 AN IDEA SHEET

MIXED CROPPING
Leafy vegetable and sunnhemp together as mixed crop

The following farmers will carry out this trial:

1. B. Yadaiah
2. Edaiah
3. B. Mahankalamma

What would we try out together?
Growing leafy vegetable and sunnhemp together and assessing how far this will be useful. A small portion of the field will be allotted to grow only the leafy vegetable and the rest will have mixed crop (leafy vegetable and sunnhemp). This will be used as a trial to compare the yield and also the profit.

What effect are we hoping for?
1. Sunnhemp hay will be available during the summer.
2. A portion of the sunnhemp can be left for seed purposes, which can either be sold or used for next season.
3. Selling of leafy vegetable will yield income.

Why is this important?
1. Feeding of sunnhemp increases the fat content in milk.
2. Animals get more protein.
3. The health of the animals is improved.

How do we implement this trial?

Who does what?

The farmers have each decided to allot the following portion of land for this purpose:

1. B. Yadaiah   – 3 guntas
2. Edaiah   – 3 guntas
3. B. Mahankalamma   – 5 guntas

The preparation of land, procuring sunnhemp and leafy vegetable seeds, spraying fertilisers etc will be done by the farmers. The expense of these activities will be borne by them.

When are which actions to be done?

Procuring seeds	– November
Land preparation	– December
Irrigating the land	– End of December
Sowing of seeds	– January

Source: Scheuermeier & Sen (1994)

LEARNING ACTIVITY 1
STUDYING FARMERS' RECENT INNOVATIONS

between 2 hrs & 2 days

Objective:
- To increase awareness of the farmers' role in generating new farming practices independent of the influence of outsiders.

Setting/approach:
- The participants go to farmers to study recent innovations on their farm.

Materials:
- Possibly a handout clarifying the assignments, prepared on the basis of the information below.

Procedure:
- The participants form teams of 2 or 3 people.
- Each team makes a visit to one or several previously selected farmers, after having been given the following instructions:
  - prepare a short list of the most important questions for interviews on recent innovations in agriculture;
  - carry out a semi-structured interview with each farmer (man or woman);
  - afterwards, prepare good, technical descriptions of a few of the innovations which farmers have developed without external support;
  - conclude your report with an estimation of which part of the farmers' innovations originated from their own efforts (including contacts with neighbours, relatives, other villages, etc) as compared to those originated by support agencies (e.g. extension services, NGOs, traders).
- If necessary, suggest that the teams include the following questions in their checklist:
  1. What changes are happening now in farming practices, compared with changes in the time of their parents?
  2. What are the most recent changes they have been involved in?
  3. How did they develop this new idea (referring especially to the source of the new idea)?
- After the teams have returned, each presents a report in a plenary session.

Variation:
- The above refers to a field visit as part of a workshop. This assignment may also be given to participants as a follow-up task after a workshop in their own working area. They may then be encouraged to use different methods to find farmers' own solutions, such as those mentioned in the discussion section of this unit, e.g. an innovators' workshop.

LEARNING ACTIVITY 2
LOOKING FOR SOLUTIONS

45–60 min

Objective:
- To become familiar with the methods most often used in searching for possible solutions to be tried out.

Setting/approach:
- A quick brainstorming session on the experiences of participants regarding sources of research options.

Materials:
- Facilities to visualise the main points of the brainstorming: a blackboard with chalk or small cards, markers, and a wall to hang the cards.

Procedure:
- Explain the objective of this exercise – for example, the need for PTD facilitators to have a good overview of where relevant ideas can be found – as well as the general brainstorming procedure.
- Ask the participants to inventorise, in pairs, methods for how or where to find possible solutions for identified agricultural problems.
- After 10 minutes, ask the pairs to report one method each and list these on the blackboard; a second, and perhaps a third, round may be necessary to collect additional methods not yet mentioned.
- In a plenary discussion, structure the results; for example, marking those methods in which the PTD facilitator plays the central role (e.g. a literature study), those where the farmers play the central role (exchange visits, etc), and those asking for a combined effort (innovators' workshops, etc).

One conclusion may be that most methods mentioned still fall into the first group: controlled by the PTD facilitator. A plenary discussion may then challenge participants as to whether or not farmers can provide relevant ideas. If so, why not try to find other ideas for farmer-controlled solutions?

Variations:
- This learning activity may be used to give one or more participants an opportunity to moderate a creative brainstorming session. Such brainstorming sessions are a crucial element in most methods used to identify possible solutions together with farmers (compare next learning activity).
- Instead of using the blackboard, one may ask participants to write each method on a card. This promotes greater activity, with participants walking to the wall to place their cards. Processing of the results, grouping similar cards, will probably take some more time and stimulate the creativeness of the facilitators.

LEARNING ACTIVITY 3
MODERATION OF A CREATIVE BRAINSTORMING SESSION

1–1½ hrs

Objective:
- To enhance participants' skills in moderating a creative brainstorming session.

Setting/approach:
- In the classroom, one or two participants are asked to moderate a brainstorming session with the other participants on a selected issue.

Materials:
- Pen and paper, blackboard and chalk.

Procedure:
- Ask two participants to make use of the opportunity to moderate the session.

- Choose an issue/subject for the brainstorming; one possibility is the subject of the previous learning activity, i.e. methods of looking for solutions. Any other subject may be chosen which is of relevance to the participants at that moment, as long as it invites them to think creatively and give ideas based on their own experiences.

- Allow the two moderators 10 minutes to prepare themselves; suggest a possible procedure for the session, such as the method used in the previous learning activity. If asked for assistance, give suggestions on how to induce a creative atmosphere during the brainstorm.

- The participant-moderators commence the brainstorming session, as prepared, and take it right up to the formulation of final conclusions.

- Evaluate the experiences of this exercise together by asking first for reactions from the moderators, then for comments from the other participants, and finally adding your own remarks.

Important issues for attention include:

- Were there participants who did not take an active part in the session? Does this mean that they did not have interesting ideas? How could we stimulate their participation next time?

- Was there an atmosphere of creativity (even strange remarks accepted, mutual stimulation rather than mutual criticism, speed of discussion maintained, attention paid to formulating new ideas rather than to details of previous ones, etc)?

- How could more creativity be stimulated next time?

- Be prepared that the subject of the discussion itself may demand some additional attention during this plenary session, if people feel strongly about the issues raised. If necessary, agree on a separate follow-up discussion of these issues so that the plenary discussion can focus on the moderation itself.

LEARNING ACTIVITY 4
ELICITING CRITERIA

1–1¼ hrs

Objective:
- To enhance participants' skills in moderating a group discussion aimed at eliciting criteria.

Setting/approach:
- The participants are given an opportunity to practise the use of a pairwise-ranking technique in a group discussion setting.

Materials:
- Blackboard and chalk or any other medium to visualise the main points of the plenary discussion.

Procedure:
- The activity follows the same basic procedure as the skill-enhancing Learning Activity 5, Unit 2.C, p.125.
- Choose a subject for pairwise ranking, preferably one that has relevance for the group of participants themselves, for example:
  - *a technical subject:* such as a comparison of several cover crops for use in intercropping;
  - *a policy subject:* in your agricultural programme, do you tend to work through individual farmers, informal farmer groups, or farmer organisations? Why?
- Work in several smaller groups for the pairwise ranking; at least one pair should be compared by a group whose main goal is to find the underlying criteria; a second pair and third pair may be done to show that this leads to new criteria.
- The plenary discussion may focus on:
  - evaluating the ranking process, the role of the moderators;
  - the difficulties of pinpointing criteria in a long argument;
  - the representativeness of the criteria thus created; for whom?
  - the feasibility of performing such an exercise together with farmers.

LEARNING ACTIVITY 5
OWN PRIORITIES

1½ hrs

Objectives:
- To increase participants' awareness of their own criteria and those of the organisation, and their relevance in PTD.
- To create an understanding of the importance of issues such as LEISA-HEIA, entry-point activity and associated support structures in selecting options for farmers' experimentation.

Setting/approach:
- A combination of small-group work and plenary discussion.

Materials:
- Newsprint and markers.

Procedure:
- Explain the objectives of the activity and the need for critical awareness of one's own criteria, and for sharing these with others.
- Ask the participants to form small groups of 3–4 people and give them the following instructions:
  - interview each other one by one on the following topic: can you mention any agricultural activities, proposed by farmers for experimentation, which you or your organisation would *not* accept to support?
  - for each example, say *why* it would not be accepted;
  - after noting the answers to these questions, per participant, use these to draw up a joint list of criteria used in accepting proposals from farmers, as identified by your group, and write the list on newsprint.

  Give examples to the groups to clarify the central topic, if necessary.

- Ask the small groups, after 30–45 minutes, to present their results briefly to the others.

- Moderate a plenary discussion on the results of the small-group work, focusing, for example, on:
  - the relevance of particular issues such as LEISA-HEIA, entry-point activity and support structures;
  - the importance of being aware that these criteria exist and the need to make them explicit;
  - the role of one's own criteria versus farmers' control of the development process via their own criteria; how to strike a balance?

LEARNING ACTIVITY 6
MAKING A CHOICE

1½–2 hrs

Objectives:
- To enhance skills in using the matrix-ranking technique as a tool in moderating decision-making.

- To promote participants' sensitiveness to the role of different actors at village level in the decision-making process.

Setting/approach:
- In the classroom, participants practise group matrix ranking with a real subject.

Materials:
- Board or large paper to prepare the matrix.

Procedure:
- The activity follows the same basic procedure as the skill-enhancing Learning Activity 5, Unit 2.C, p.125; the matrix ranking may be moderated in two parallel groups to give more participants an opportunity to develop their moderation skills.

- To select a subject for the matrix ranking, the results of one of the previous exercises may be used, in which lists of options versus lists of criteria have already been generated, for example:
  - technical options (e.g. cover crops) versus the criteria identified in Learning Activity 4 (eliciting criteria, p.137); or
  - the criteria of one's own organisation, as developed in the previous activity, versus a set of possible research options.

- The plenary discussion generally has two parts:
  - first, an evaluation of the ranking process, the effectiveness of the moderators; how preferences were indicated and the scoring or ranking process.
  - second, an assessment of the feasibility of using this tool in decision-making at a village/farmer level: how realistic is the decision-making process suggested through the exercise? the (decisive?) role of formal and informal leaders? how to involve the poor, women, young farmers etc in decision-making? the advantages and danger of the PTD team manipulating this decision-making process?

2 / Towards an Agenda for Action • D / Looking for Things to Try

LEARNING ACTIVITY 7
HYPOTHESIS FORMULATION

45 min

Objective:
- To enhance participants' skills in formulating a hypothesis.

Setting/approach:
- The participants are given an opportunity to try the formulation of a clear hypothesis individually.

Materials:
- Pen and paper, blackboard and chalk to note the results of the exercise; possibly an overhead projector.

Procedure:
- Present to the participants a very brief case describing a certain problem together with a possible solution for this problem.

  In selecting or preparing a case for this exercise you may consider the following:
  1. for a first exercise, select a simple problem and solution; a second round may study a more complex situation;
  2. the simple problem/solution may be derived from agriculture, but an example from day-to-day life is often more revealing; for example (for male participants): if I always wear a hat, I will have more success with the opposite sex, because my baldness will remain unnoticed;
  3. the zero-grazing example in Box 2.12 (p.132) may be used to develop a more complex case;
  4. but often previous sessions on problem diagnosis and identification of solutions give enough material for a case in this exercise.

  The case may be presented on an overhead projector or in the form of a handout.

- Ask the participants individually to formulate the presented solution in the form of a hypothesis (if…, then…, because…).

- Ask one participant to report his or her hypothesis and note this on the blackboard. Then ask other participants to provide hypotheses that are slight variations on the first one. Finally, ask participants to provide hypotheses that are quite different from the first one; note variations on the blackboard.

- Facilitate a plenary discussion on the merits of the different hypotheses, paying special attention to the need:
  - to be precise; and
  - to indicate clearly cause–effect relationships.

- You may conclude this session with a general discussion on the role of good hypotheses in the PTD process.

FARMERS' EXPERIMENTATION

PART 3

PART 3 / FARMERS' EXPERIMENTATION

3·A / LOOKING AT LOCAL WAYS OF EXPERIMENTATION

- Experimentation by farmers — 145
- Comparing farmers' and outsiders' experiments — 146
- Identifying local experimenters — 147
- Farmers' approach to experimentation — 148
- Learning about farmers' experimentation — 149
- Giving value to local experimentation — 150

3·B / STRENGTHENING FARMERS' EXPERIMENTATION

- Strategies to support local experimentation — 156
- Basics of systematic experimentation — 157
- Improving farmers' experimental design — 157
- Farmer experimental design workshops — 158
- Going deeper into trial formats — 160
- Four warnings — 160
- Stepwise testing to develop farm systems — 161

3·C / FARMER-EXPERIMENTER GROUPS

- The role of farmer-experimenter groups in PTD — 170
- Formation of farmer-experimenter groups — 171
- Characteristics of successful self-organisation — 172
- Planning group meetings — 174

3·D / RECORDING AND ASSESSING EXPERIMENTS

- Importance of recording and assessment — 176
- Planning a recording and assessment system — 177
- Farmers' criteria — 177
- Collection and recording — 178
- Compilation and analysis — 179
- Comparing "costs" and "benefits" — 180
- Effects on the farming system — 180
- Preparing for the next round of experiments — 181

BOXES	3.1	Reasons for promoting farmers' experimentation	146
	3.2	Limits to farmers' experimentation	147
	3.3	A comparison of scientists' and farmers' research	148
	3.4	Design issues in improving farmers' experimentation in cropping	158
	3.5	Case study of farmers' experiments on germination of teak seeds	163
	3.6	Cases of farmers' experimentation	167
	3.7	Mr Casas's farm: an example of gradual transition to ecological farming	169
	3.8	Key questions in developing a recording and assessment system	177
	3.9	Case study: "No milk for the dairy plant?" – additional information for trainers	186

TABLES	3.1	Group development process: experiences from Sri Lanka	173
	3.2	Summary of results of experiments on survival rates	179
	3.3	Treatment comparison table	179

HANDOUTS	3.1	Farmers' experimentation: assignment for field study	153
	3.2	"Malika Fernando is ill"	182
	3.3	Elaboration of exercise "Malika Fernando is ill"	183
	3.4	Case study: "No milk for the dairy plant?"	185
	3.5	Case study: farmer group "Joining Hands"	188

UNIT 3·A / LOOKING AT LOCAL WAYS OF EXPERIMENTATION

OVERVIEW OF THIS UNIT

EXPECTED RESULTS

This unit is designed to develop fieldworkers' insight and skill in recognising how farmers carry out their own experiments, and to enhance the value that farmers and outsiders (particularly researchers and extensionists) attach to farmers' experimentation.

After completing the learning activities in this unit, the participants are expected:

- to understand the social basis of knowledge generation and the importance of building on farmers' experimental capacities and methods;
- to be able to identify farmers with above-average experimental skills and practice;
- to discuss and systematise with farmers their ways of experimenting and innovating in agriculture, and to appreciate the logic behind it;
- to be able to deal with the contradictions between their own and farmers' experimental logic and designs.

MAIN CONCEPTS

- **Farmers' experimentation:** This is farmers' capacity to change their situation through experimentation and refers to experiments that farmers define and control themselves, making their own observations and analysis.
- **Local experimenters:** This refers to the individuals in the community who develop or try out new ways of doing things.
- **Farmers' methods and logic of experimentation:** The "how" and "why" of farmers' experimentation.
- **Cultural dimension of experimentation:** How farmers and communities give meaning to their environment through experimentation.

TRAINING METHODOLOGY

A good start can be made by gathering what participants already know about innovations developed by farmers in their area. Some participants may find this difficult at first, but as soon as a few examples are given, others will quickly follow. If there are participants who have their own farms or have a farming background, the process of experimentation and innovation by these individuals can be studied in more detail. In most cases, however, direct interaction with farmers in the field will provide the most intensive learning experiences.

LEARNING ACTIVITIES

1. Inventory of farmers' experiments
2. Field study of farmers' experimentation
3. Comparing how farmers and outsiders perceive "innovations"
4. Producing media to document and spread local innovations
5. Inventory of local measurements and calculations

DISCUSSION

EXPERIMENTATION BY FARMERS

Farmers are continuously experimenting. Trying out new things is an essential part of farming. Modern agricultural science rests on the foundation of millennia of informal experimentation. Farmers carry out experiments not only in reaction to outside influences, such as the introduction of new technologies by extension agents, but also on their own initiative.

In Mali, numerous farmers' experiments were observed (Stolzenbach, 1993), e.g.:

- A farmer in Sanando obtained a new variety of beans and planted them in a corner of his field to see whether they performed better than his own variety.
- Another Sanando farmer tried out the effect of using fertiliser on groundnuts, whereas others used it only for cotton. He was satisfied with the increase in yield but discontinued the practice because the groundnuts had a disappointing taste.
- A farmer from Koyan, who had started to grow a shorter millet variety, reduced the sowing distance gradually over several seasons until he found the optimal distance.

In Kenya, a woman farmer has developed a practice of *"boma-*mulching" through her own observation and experimentation. She trims branches from living fences of *Euphorbia terucalli* and spreads them in the *boma* (cattle pen), where they get soaked with manure, trampled by the animals, and worked into the manure and soil. This quickly produces a kind of compost that she uses to improve soil fertility in her fields (Chambers et al., 1989). Another Kenyan farmer conducted her own trial using low levels of fertiliser application on a clean-tilled plot with millet and cassava, with a partial control (clean-tilled, no fertiliser). This trial combined the land-preparation techniques for monocropped maize with lower fertiliser levels and with food crops that are less risky than maize (Rocheleau, 1988). Similar examples of informal experimentation by farmers are found in Box (1987), Brouwers (1993), PMHE (1992) and Richards (1986).

Farmers carry out experiments in almost all aspects of farming, with a certain emphasis on crop and livestock breed selection, animal feeding and care, crop protection, fertilisation and other cultural practices in cropping, and the processing and storage of crop and livestock products.

Farmers may conduct these experiments for a number of reasons (Rhoades & Bebbington, 1991):

- out of curiosity – just to try out an idea that comes to mind;

- to solve a problem – to find solutions for current pressing problems;

- to adapt technologies to local conditions and to the farmers' specific interests and preferences.

They conduct adaptation experiments when they want to test an unknown technology in a known environment, or a known technology in an unknown environment, such as after resettlement.

COMPARING FARMERS' AND OUTSIDERS' EXPERIMENTS

Not only research organisations but also many development projects and NGOs are engaged in some type of agricultural experimentation. This may involve trials in their own fields ("on-station" research) or in farmers' fields ("on-farm" research). However, most of these trials are controlled by the staff rather than by the farmers for whom the innovations are intended. Research organisations in particular try to keep their experiments under their own control. The staff of the project or organisation may refer to "participatory research" or even "farmers' research", even though the farmers' participation is limited to providing land and labour or to commenting on trial results.

Projects and NGOs become involved in agricultural research for various reasons, such as:

- government research programmes do not include the type of technologies that small-scale farmers are interested in;

- research was not done in the agro-ecological zone where the project or NGO is operating;

- the staff want to develop their own understanding of certain technologies and develop essential skills before starting to work on these technologies together with farmers;

- the staff want to screen new technologies to eliminate any risky options, before suggesting the innovations to farmers;

- the organisation wants to show that it is capable of carrying out serious quantitative research.

Whatever the reason may be, such research activities by projects and NGOs must be distinguished clearly from farmers' experimentation, which refers only to

---

BOX 3-1 REASONS FOR PROMOTING FARMERS' EXPERIMENTATION

- Because we are rarely sure that the solutions we promote fit the biophysical and socio-economic conditions of the local farmers.

- Because the contrast between on-station and local farming practices is so great that farmers, especially women, are alienated.

- Because there is no way that formal research and extension can provide answers to the diverse, site-specific problems and challenges faced by farmers (how many researchers are there for how many farmers?).

- Because it is an effective tool for step-by-step development of local agriculture.

- Because it strengthens farmers' confidence and capacity to solve their own problems and reduces their dependency on outsiders.

- Because farmers do it anyway, and neglecting it would be completely contrary to a participatory approach to agricultural development.

Source: adapted from PMHE (1993)

experiments that are defined, controlled, implemented and assessed by the farmers themselves, using their own inputs and doing their own observation and recording. PTD gives priority to supporting this type of experimentation. Important reasons for doing so have been listed in a training session in Sri Lanka (Box 3.1).

Farmers' experimentation faces certain limits (Box 3.2). It is complementary to research controlled by scientists (Box 3.3). Controlled research is required when high risks are involved, complicated equipment is needed, results that can be interpreted can be expected only after several years, or control of variables is crucial for assessing the results. Supporting farmers' experimentation through a process of PTD is required when technologies specific to certain sites or farming systems are being sought. As will become evident in the pages that follow, this process can lead not only to appropriate technologies but also to increased capacity to innovate, among farmers and scientists alike.

IDENTIFYING LOCAL EXPERIMENTERS

To find out how farmers conduct experiments, it is best to start by identifying local men and women who are active experimenters. To a certain extent, all farmers carry out experiments. However, the term "local experimenters" refers to those who are known in their community as innovative people. More than the average farmer does, they develop new ways of doing things. On their own initiative, they try out new varieties, crops, breeds or species of animals. They are generally quicker to perceive new opportunities. By observing and discussing with these people, we can discover the logic behind their experiments, the experimental methods they use, and the strengths and weaknesses of these methods.

Local experimenters should not be confused with those people, sometimes called "progressive farmers", who adopt introduced technologies because they have ample resources, intensive contact with extension agents, and easy access to other external resources and services. Their situation differs from that of less well-endowed farmers. There is some basis in the saying that "necessity is the mother of invention". Farmers without the means to adopt high-external-input technologies but under pressure to prevent their situation from deteriorating can be very innovative in experimenting with "unconventional" inputs and techniques.

BOX 3-2 LIMITS TO FARMERS' EXPERIMENTATION

When farmers perceive problems such as decreasing levels of soil fertility or increasing soil erosion, they may try out various potential solutions, either conceived by them or known to them. However, the errors in trial-and-error experimentation may be costly, particularly in terms of time. In their responses to problems or opportunities, farmers are not aware of all the possibilities developed outside their communication network, nor can they be aware of all the repercussions of new technologies. Scientists, too, are not fully aware of all the possibilities and repercussions, but they may have more systematic and wider-ranging methods of recognising them.

In his observations of spontaneous experimentation by farmers with a new crop in Thailand, Connell (1990) noted the following limitations to the farmers' abilities and effectiveness in generating new technologies.

- **Undirected experimentation.** In their enthusiasm to experiment with the new technology, farmers liked to think up their own personal variants. Other technology variations occurred by chance, without the farmers being aware that they were doing something different from their neighbours.

- **Lack of analytical approach.** Many of the farmers were not analytical in evaluating the techniques they tried in their fields and were in danger of coming to false conclusions. They did not always understand the underlying reasons for a good or poor yield and attributed the success of a technique to the most obvious difference. For example, in one village farmers compared wheat plots on the basis of whether they were broadcast or row-seeded, when the main reason for the varying stands was the extent of over-irrigation.

- **Poor experimental design.** Experimenting farmers sometimes did not design comparable units. When they tried out a new technique, the basis for comparison was a previous season's yield or crops in a nearby field. Again this may lead to false conclusions because of different soil types or management system that would invalidate the conclusion.

Connell concluded that, for these reasons, the outcome of farmers' technology development is undirected and uncertain. These are areas where farmers' abilities could be strengthened and developed. However, these limitations do not invalidate the concept of farmers' experimentation. While the experiment by any one farmer might not be productive, it is very likely that some worthwhile innovation will be developed by farmers when the process takes place within a farming community or larger population with well-functioning (informal) communication channels.

Source: Reijntjes et al. (1992)

BOX 3-3 A COMPARISON OF SCIENTISTS' AND FARMERS' RESEARCH

SCIENTISTS' RESEARCH	FARMERS' RESEARCH
Capital-intensive equipment	Limited equipment, use of local resources
Long-term perspective	Short-term perspective
Complicated design and analysis	Farmer-determined design and analysis
Standard procedures	Non-standard procedures, according to *ad hoc* needs
Site-specific	Site-specific
Chemical means	Natural means
Single commodities	Integrated systems
Oriented to (urban) consumer preferences and markets	Oriented to local preferences and markets
Controlled variables	Follows farmers' management practices
Artificial situation	Real-life situation

Local experimenters can be identified by:

- **Observation:** Walk slowly through the area where you work and take a close look at the farmhouses and fields: where do you see unusual things, variations in the common pattern of farming and living?

- **Chain interviews:** Ask some key informants to mention names of persons (better-off, average and poor; male and female) who are very creative and like to try out new things; visit these farmers, interview them about their experiments and ask the names of other "experimenters" they know.

- **Stepwise selection:** For a certain crop or animal or activity, first ask farmers or key informants for the names of farmers whom they consider to be the local "experts" – people who have long-term experience and above-average knowledge and skills in that field. In many cases, these experts will be women, especially (but not only) when food crops, small animals, food processing and storage are concerned. Interview those people considered to be most creative and innovative. Differentiation can also be made between farmers who experiment mainly with introduced technologies and those who experiment mainly with local technologies.

- **Reconstructing innovations:** Ask a group of farmers to list one or more agricultural innovations that have been made in the last ten years and are relevant for most of the families in the area. Ask them to identify the farmers who played an important role in introducing, adapting or developing these innovations.

FARMERS' APPROACH TO EXPERIMENTATION

Adapting technologies to local conditions and preferences is not just a technical and economic process. Farmers' experimentation is also a process of appropriation ("making it one's own") by transforming the technologies coming from outside the community and synthesising them with the local culture. Farmers' experimentation is closely related to the cultural concepts and values of their social group: their ways of thinking and communicating, their relationship to nature, the norms that shape their social organisation.

For example, knowledge generation by farmers in the mountains of Peru involves the use of intuitive methods and particular cultural conventions and idioms. It is strongly related to the spiritual and social dimensions of the community; their concept of the cosmos and the strong reciprocity in relationships between family and community that structure their perception, thinking and action (Salas *et al.*, 1989).

In Ghana, the way the Talensi see the cosmos – the interaction between man, nature and the gods – leads them to analyse the occurrence of events and agricultural change in an interrelated way. This viewpoint is contrasted to the reductionist approach to problem-solving which is often practised by Western-educated outsiders. An experiment in maize, for instance, may be considered a failure, despite a perceptible yield increase, if a death occurs in the family. Because of the crucial role played by local soothsayers and other indigenous institutions in the analysis of such events,

PTD practitioners should link up with them in experimental activities (Haverkort & Millar, 1992).

Farmers' experimental practice often does not comply with researchers' criteria for systematic experiments. In the Sanando area of Mali, Stolzenbach (1992 and 1993) noted the following about farmers' experimental methods and logic:

- Experiments were often not isolated events planned in advance but part of everyday farming; one can speak of a continuous innovative process.

- Farmers often learned from spontaneous experimental events. For example, when children and adults sow beans together, young children with short legs may sow with less distance between seedholes than in the rows sown by adults. Farmers, who observe the resulting differences in plant performance, then learn from this.

- When carrying out planned experiments, Sanando farmers:
  - waited until the main crop(s) were in the fields, in order to safeguard their subsistence;
  - preferred to do the experiment in their main fields;
  - preferred to allocate a large part of their fields to an experiment – as one farmer put it: "an insignificant plot will give insignificant results" (Stolzenbach, 1992); but, if little seed was available or certain risks were foreseen, they opted for a small plot;
  - analysed in detail those experiments that showed new ideas to be ineffective, in order to identify circumstances that may have influenced the outcome, but rarely analysed experiments with positive results in similar detail;
  - were often satisfied with single-season experiments; they repeated experiments during a second season only when first-season results were unexpectedly negative;
  - carried out mainly comparative studies, in which only one variable differed between the treatments;
  - were, however, aware that influences could also be exerted by other variables and external factors such as the amount and timing of rain or the occurrence of certain pests. When assessing the results of the experiment, they combined all this information intuitively. Long-term experience with their farms played an important role in this assessment.

LEARNING ABOUT FARMERS' EXPERIMENTATION

In each specific situation, the locally relevant aspects of farmers' experimental methods and logic must be learned by PTD practitioners. In this process, eyes and ears should be kept open for the related perceptions, norms and interests of the farmers. Care must be taken not to isolate farmers' experimental methods from their socio-cultural basis. Learning about farmers' ways of experimenting is learning about farmers' way of life.

Numerous groups of questions need to be explored when trying to learn about how men and women farmers carry out their own experiments. These include the following:

- **Justification:** Why do farmers decide to experiment? Do they react to actions by other farmers, external incentives, changing situation, or something else? Are women's reasons for experimenting different from men's?

- **Planning:** How do farmers plan their experiments? To what extent are these consciously and systematically prepared? How are tasks and responsibilities related to planning and managing certain types of experiments divided according to gender? How many experiments are farmers undertaking simultaneously at individual, family or community level?

Experiments are often part of everyday farming

- **Hypotheses:** How are research questions formulated? How do the research questions reflect how farmers think about nature, farming and innovation?

- **Variables and levels:** How many variables and treatment levels are included? If complex innovations are tested, how do farmers manage to monitor and assess the contribution of the various components to the final result?

- **Non-treatment or control:** With what do farmers compare the new practice or technology? Do they use "control plots" or "control animals"? If not, how do they take external influences into account when interpreting the results?

- **Layout and timing:** How do farmers locate the trials in their farm and in a field? How big are the experimental plots? Do they select only certain animals for a trial and, if so, what type of animals? What is the timing for the beginning and end of a trial? For how many years do they repeat a trial before drawing definite conclusions? Why do the experimenting farmers make these choices? Does this vary with the type of experiment concerned? How is the design of a trial influenced by the gender and age distribution of tasks and responsibilities within the family (think of "women's" crops or fields, or animals to which women have the rights to the products)?

- **Data collection:** How do farmers gather information during and at the end of the experiment? What does the farmer observe and measure? When and how are these observations made? How do farmers keep track of what was done and what resulted? What units of measurement and which type of "records" do they use?

- **Analysis:** What criteria do farmers use when assessing the results of their experiments? What is the relationship between these criteria and the specific tasks and responsibilities of the farmers (e.g. food production, processing, sale)? Do men and women exchange information and views between themselves? In learning about farmers' experimentation, similar methods and tools can be employed as were used for identifying local experimenters: observations, chain interviews, stepwise selection and reconstructing innovations. In order to prepare for this, it is enlightening to study first the words and expressions in the local language that are used for "trying out" or "experimenting". In what other contexts are these same terms used? What is their direct translation back into English? What does this say about how farmers perceive experiments?

GIVING VALUE TO LOCAL EXPERIMENTATION

PTD is primarily aimed at stimulating the generation of local knowledge and reinforcing local capacities to develop sustainable farming systems. Facilitators encourage farmers and farm communities to analyse how they carry out experiments, to relate this to their way of life (their culture), and to recognise the value of their own experimentation and innovation.

Initially, farmers often react reluctantly when outsiders refer to the possibility of doing experiments together. This may be due to a misunderstanding about what is actually referred to but it can be overcome by using the local words and expressions for "trying out". There may also be a justified fear of becoming involved in an overly risky adventure.

In many cases, the pressure for "modernisation" and the status given to formal "scientific" research has made fieldworkers and farmers believe that farmers' experiments are relics of a traditional way of farming that must be replaced by scientific methods. Local creation and sharing of knowledge have often been denied recognition, and farmer experimentation has rarely been valued. This has been the case especially with food crops or "women's crops". In such cases, outsiders can play a role in helping farmers to recognise the value of their informal experimental activities and in restoring their confidence in their own potential to improve their farming.

Possible ways of stimulating farmers to recognise the value of their own experimentation include:

- **Storytelling:** encourage farmers to tell stories, sing songs or give performances (e.g. dance) about important innovations made by farmers in the past;

- **Reconstructing innovations:** together with a group of farmers, analyse the origin of the most significant improvements in their farming systems and farmers' contributions in introducing those improvements;

- **Producing local media:** encourage documentation and the sharing of recent cases of farmers' experimentation and innovation in songs, poems, proverbs, plays, etc, or by using "modern" media such as farmer-made photographs or drawings, posters or cassette recordings;

- **Local education:** encourage farmer-experimenters to meet with village youth and schoolchildren to tell (and show) them about their experimentation, how they do it and why they do it in that way. The young people may, in return, assist in documenting and disseminating these experiences and skills;

- **Socialisation:** encourage farmers to incorporate the foregoing activities into community institutions for communication and learning: for example, as part of community celebrations, in school programmes, or as a regular feature of meetings of local farmers' organisations.

A key element in all these activities is the genuine interest shown by outsiders in farmers' own initiatives and efforts. Encouraging visitors from elsewhere, either officials or other farmers, to see and discuss these activities can greatly reinforce this element.

LEARNING ACTIVITY 1
INVENTORY OF FARMERS' EXPERIMENTS

2 hrs

Objective:
- To make participants aware of ongoing experimentation by farmers and its importance for agricultural innovation.

Setting/approach:
- Basically a classroom situation to generate cases of farmers' experimentation for further analysis and discussion.

Materials:
- Cards and markers, tape, and a large board or wall for attaching cards.

Procedure:
- Ask participants, in pairs, to identify some cases of farmers trying out improvements on their own initiative, asking:

  • With what was the farmer experimenting?
  • Why was the farmer experimenting?

  For example: one farmer tried a new vegetable variety, because it was given to her by a friend and she wanted to see how it grew.

- Ask the pairs to note each experiment on a card and hang all cards on the board/wall.

- Go through the cards together and ask participants to clarify those that are not clear.

- Moderate a short plenary discussion asking the participants to cluster cards referring to related or similar experiments. Clusters may develop around breed selection, fertility management, pest and disease control, feeding practices, cultivation practices, processing, storage, etc.

- Moderate a discussion of the results obtained, focusing on:

  • the number of examples found, and conclusions about the extent to which farmers' experimentation occurs;
  • the most common topics of experiments;
  • the reasons why farmers experiment (participants' remarks can be compared with the three reasons given earlier by Rhoades & Bebbington, 1991).

This plenary discussion may reveal some confusion about what exactly is meant by farmers' experimentation, especially as compared with farmers' indigenous practices, and about the difference between an innovation and an experiment. This should lead to an increased understanding that experimentation is a purposeful activity, part of active learning by farmers; it is not just doing something in a different way.

Variation:
- The same process using cards may be repeated on the question: "What are strong and weak elements/aspects in farmers' own experimentation efforts?" A summary of the weaknesses in the plenary will provide a good basis for thinking about ways to support and strengthen farmers' experimentation.

LEARNING ACTIVITY 2
FIELD STUDY OF FARMERS' EXPERIMENTATION

6–10 hrs

Objectives:
- To make participants aware of ongoing experimentation by farmers and its importance for agricultural innovation.
- To make participants aware of possible differences in experimentation by men and women farmers and the reasons behind these differences (division of labour, access to or control over resources, etc).
- To reflect on how farmers experiment and the reasoning behind it.
- To strengthen participants' skills in finding out about farmers' logic and ways of experimentation.

Setting/approach:
- Participants study farmer experimentation in the field and interact in small groups with local experimenters.

Materials:
- Handout 3.1 with terms of reference for fieldwork.

Procedure:
- The facilitator explains the objectives and procedure, stressing that the emphasis will be on describing farmers' experimentation as it is actually done, without worrying about how it "should" be done according to what is written in books.
- The participants form small groups and study the handout for the fieldwork. They prepare themselves by agreeing on the main points of attention (checklist) during fieldwork, and the ways of finding out about farmers' experimentation (identifying experimenters, the terms used, and the division of tasks within the team).
- In a plenary session, for mutual feedback and learning, the small groups report the main outcome of their preparations to each other. The groups can be given the handout, listing some important issues in learning about farmers' experimentation, to compare with their own checklist.
- The small groups go into the field and study with two or more men and women farmers how they experiment and why they do it this way. The emphasis should be on in-depth studies of a few experiments rather than collecting a great number of examples in a superficial way. The cases are documented (the process as well as the results of the experiments), and can be used as future training materials. Before going to the field, participants may discuss and act out a role play about how to talk about such matters with farmers (see also the learning activities on interacting with farmers in Units 1.G, pp. 79–83, and 2.C, pp. 121–8).
- After returning from the field, the small groups prepare for reporting the main outcome of their work to the other groups. The plenary discussion of the results could focus on:
  - common logic and methods of farmers' experiments;
  - differences between these and the logic and methods of conventional scientific experiments;
  - possible differences between men and women farmer-experimenters;
  - possible shortcomings in farmers' experimental methods;
  - techniques that fieldworkers can use to find out about how farmers experiment.

During the plenary discussion, it may be helpful to present an overview of the main characteristics of scientists' vs farmers' research (see Boxes 3.2 and 3.3).

Variations:
- If there is no opportunity for fieldwork, individual participants who are still active in farming themselves (preferably coming from a farming family in the working area) can be the "farmers" who are interviewed by the small groups.
- The comparison of scientists' and farmers' research may arise in the plenary discussion but can be referred to a subsequent small-group discussion to allow for analysis and discussion in greater depth.

HANDOUT 3-1

FARMERS' EXPERIMENTATION

Assignment for Field Study

The field study should lead to answers to the following questions:

- What are the local words for "trying out" and "experimenting"?

- Which farmers in the area have, to a greater or lesser extent, been experimenting on their own? What led them to experiment?

- Which experiments achieved by farmers without outside help can be identified? Analyse a few selected ones in great detail.

Suggestions for fieldwork:

- You may wish to use semi-structured interviewing techniques, combined with field observations and any other PRA techniques you find appropriate.

- In preparing yourselves for the fieldwork as a team, consider:
  - division of tasks and responsibilities;
  - drafting a checklist of main questions/issues.

- Internal debriefing – i.e. collecting and compiling the information from all team members immediately after completing the interviews and observations, possibly while still in the field, to prepare for presentation in the plenary session.

Source: PMHE (1992)

LEARNING ACTIVITY 3
COMPARING HOW FARMERS AND OUTSIDERS PERCEIVE "INNOVATIONS"

6–8 hrs

Objectives:
- To make participants aware that most innovations of importance to farmers have not been "delivered" by formal research and extension systems and that, even where this is the case, they have spread only after considerable adaptation by farmers to make the innovations "fit" into their farms.

- To make participants aware of the differences between social/gender groups in the perception of innovations.

- To enhance participants' skills in identifying local innovations and experimental processes.

Setting/approach:
- An important part of the activity is in the field.

Materials:
- Handout with terms of reference for fieldwork (optional, to be prepared).

Procedure:
- Ask participants to make a list of what they perceive as important innovations made in the farming systems in their working area during the last 20 years (first individually, then draw up a group list by ranking).

- Participants form small groups and visit several farmers (especially smallholders, both male and female, and of different socio-economic, ethnic or religious groups) to find out what they regard as important innovations in their farming systems. The participants should remember that farmers often use different words for changes introduced from "official sources" and innovations made by themselves. Finding out the proper local words for these different types of changes would be a first step for the field teams.

- For one or two of those innovations, the small groups try to reconstruct with farmers how this came to be:
  - What exactly was the original idea?
  - When did the idea arise?
  - What was the origin of the idea? Who invented it or brought it to the community?
  - Who were initially involved in trying out the idea?
  - Why did they want to test this idea? What changes did they make to the original idea?
  - How did more people get to know about the innovation? Who else started using the (adapted) idea?

- The cases are documented (and can be used in future training events).

- The participants compare their own initial list with the farmers' list and discuss the differences between them and the reasons for these differences.

- Participants present and compare the "reconstructed" innovations. Then they compare their findings with commonly-held assumptions about the process of agricultural innovation: formal research as a source of new ideas, experimentation and testing by research, spreading the news by extension services, adoption of complete (unchanged) packages by farmers.

Variations:
- Instead of fieldwork, a local case study can be prepared by the trainers or taken from literature.

- Participants may be asked, at the end of the activity, to analyse how they discussed these things with farmers (what worked well and what not? why?) and to try to derive some "dos and don'ts" for finding out about farmer experimentation and innovation.

LEARNING ACTIVITY 4
PRODUCING MEDIA TO DOCUMENT AND SPREAD LOCAL INNOVATIONS

4–6 hrs

Objectives:
- To discover the importance of giving renewed social value to farmers' experimentation and innovation.
- To broaden participants' view on the "transfer of innovations" by focusing on farmer-to-farmer transfer.
- To discover the importance of "fun" in knowledge transfer.

Setting/approach:
- A classroom activity encouraging the creativity of participants. In certain cases, the results of this activity may be presented to a wider audience; for enjoyment or to increase the learning effect.

Procedure:
- Participants are asked to list some improvements made by farmers in their local farming systems. This information may be the result of earlier learning activities.
- Participants describe important cultural means of expression and communication in the culture to which these farmers belong (if relevant, this is differentiated according to gender). If participants are not from the same area or socio-cultural group, they can each do this for their own culture.
- Participants select one innovation (or way of experimenting) and describe it in a song, poem, short play, proverb, joke or any other local means of expression and communication. If participants are from a different socio-cultural group from the farmers, they can use elements of their own culture.
- Participants perform their results, and discuss:
  - the process they have gone through, including their views on the importance of enhancing the social value given to local experimentation and sharing the results of local experiments ("horizontal" transfer of innovations);
  - possible ways to stimulate such re-valuing of local experimentation in their working area.

Variation:
- If participants have only a little knowledge of the local culture or are hesitant to start producing local media, a mini-workshop could be organised, with some local artists invited to perform and assist in the process of producing local media.

LEARNING ACTIVITY 5
INVENTORY OF LOCAL MEASUREMENTS AND CALCULATIONS

4 hrs

Objectives:
- To familiarise participants with local ways of measuring, calculating and recording.
- To reflect on implications for fieldworkers' assistance to farmers in designing, recording and assessing their experiments.

Setting/approach:
- Outside the classroom, in farmers' fields, but also in kitchens, on the market, in a local workshop, etc.

Materials:
- Pen and paper.

Procedure:
- Participants together make a list of various locations or activities in which some type of measuring, calculating and recording is done: in the kitchen (preparing/processing food); in the homestead (processing, storage, small animal production); in the field, pasture or forest; in workshops of local artisans; on the market, etc.
- Participants divide up into groups, and each goes to one of the locations to observe and discuss local ways of measurement and calculation related to specific activities at that site.
- Participants present their findings to each other (document the cases for future use!) and discuss the following questions:
  - Do local techniques and units of measurement and calculation also represent other ways of perceiving and valuing things?
  - What consequences do these findings have for the support given by fieldworkers in designing and recording farmers' experiments?

UNIT 3·B / STRENGTHENING FARMERS' EXPERIMENTATION

OVERVIEW OF THIS UNIT

EXPECTED RESULTS

This unit is intended to increase fieldworkers' understanding of different options for linking with and supporting farmers' experimentation. Fieldworkers will be equipped with skills to help farmers to design experiments that are both practical and effective, and that build on local experimental practices. After completing the learning activities in this unit, participants are expected:

- to be aware of some basic principles of systematic experimentation and important considerations in designing farmer experiments;
- to be able to facilitate farmers' discussion and decision-making about designing and organising their experiments.

MAIN CONCEPTS

- **Learning strategy:** Outsiders support farmer experimentation by learning from it, documenting it, and sharing its results.

- **Add-options strategy:** The effectivity of farmer experimentation is increased by providing farmers with a deeper understanding of the non-visible processes related to technologies tested and/or a wider choice of technologies with which to experiment.

- **Improved-design strategy:** PTD staff support farmers in making their experiments more systematic.

- **Systematic experimentation:** Experiments are set up taking into consideration issues such as control, replication, border effects, number of variables per treatment.

- **Stepwise testing** for farming system development.

TRAINING METHODOLOGY

The crucial issue is finding a balance between farmers' own ways of experimenting and possible inputs of outsiders to strengthen them. Participants with a scientific and/or research background may have to be challenged on the appropriateness of their preconceptions about "proper" experimentation. Extensionists often need to develop understanding and confidence in this relatively new type of activity. Case studies provide an important means of stimulating reflection on how to support farmer experimentation. Several are given in the learning activities, but cases based on local experiences are preferable.

LEARNING ACTIVITIES

1. Promotion of farmers' experimentation: a case study
2. The minimum-requirements question
3. The old lady who once had a pain: a story highlighting the need for systematic experimentation
4. Developing guidelines for designing experiments with farmers
5. Simulation: "Farmer design workshop"
6. Mutual consultation about self-designed experiments
7. Reconstructing farm development

DISCUSSION

STRATEGIES TO SUPPORT LOCAL EXPERIMENTATION

The main question here is: what can outsiders contribute to farmers' own experimental efforts? Three basic strategies deserve to be reflected upon before any PTD programme is started:

Learning strategy: Efforts to strengthen or improve farmer experimentation can disturb processes that farmers have managed effectively for decades (Gnagi, 1992). In many cases, extensionists can play an important role just by learning from farmers' ongoing experimental activities and making these known to farmers in other villages. Extensionists can bring farmer-experimenters together to provide learning opportunities for them. Researchers can carefully document farmers' experiments and the adaptations farmers make to technologies promoted by extensionists. The understanding thus developed can also help researchers to improve their on-station work (Okali et al., 1992).

Add-options strategy: Farmers' experimental work may become more effective if they have a wider range of options and ideas from which to choose. Specific areas can be identified where farmers' lack of knowledge hinders them from finding appropriate solutions. For example, when farmers in Honduras were taught what they did not know about insect life-cycles, they were in a better position to develop alternatives to chemical pesticides (Bentley, 1992).

Improved-design strategy: After outsiders have studied the existing experimental approaches and methods together with the experimenting farmers, agreement may be reached on how to improve these and work towards more systematic forms of experimentation. Examples of improvements often agreed upon in field programmes are: limiting the number of variables, selecting controls, and demarcating test plots.

In this third strategy, fieldworkers play two crucial roles:

1. *Enhancing farmers' understanding of the basic principles behind locally developed or introduced technologies and the underlying biological processes.* Transferring "recipes", the usual extension approach, does not strengthen farmers' capacity to experiment. Farmers need to understand the principles behind a technology so that they can use them to adapt it, or to tackle new problems that arise. This implies that designing experiments is not confined, for example, to defining treatments, or layout. It should be a joint search to produce a better understanding of the biological processes involved. For example, conventional fertiliser demonstrations are concerned with the type and amount of fertiliser to be applied, whereas a PTD experiment is concerned with how different types of fertiliser affect soil life, fertility, water retention or other related aspects;

2. *Facilitating systematic discussion and decision-making among interested farmers about how to design their experiments.* This will be the focus of the rest of this unit. Such discussions and the practical experience in collaborative experimentation should help farmers to understand better the basic principles of systematic experimentation. Improvements in ways of experimenting should be presented as options for farmers to consider. Otherwise, one falls into the old trap of offering a standard recipe – in this case, not a technology but a "package" of systematic experimentation methodologies.

BASICS OF SYSTEMATIC EXPERIMENTATION

If a choice is made to improve experimentation practices, three basic requirements for achieving this are:

- **Clear hypotheses/objectives.** It is crucial to be clear on what is tested and why. The hypothesis expresses an expectation about the cause–effect relationship between the variables involved in the experiment: "We expect that, if we do this, such and such will happen." The hypothesis need not be meaningful for the assisting outsider. For example, for a given crop variety a farmer may want to compare the sowing date recommended by the extensionist with the sowing date culturally defined by the cosmological calendar.

- **Replicable testing procedures.** It should be clear to everyone how the hypothesis is to be tested and under what conditions the experiment is to be carried out. Farmers should be able to specify what conditions will strongly influence the outcome of the experiment and up to what point, as well as how they want to control or measure these conditions. Farmers' and outsiders' views may differ as to which conditions should be taken into account and controlled. This is often why trials designed by outsiders (even on-farm trials implemented by farmers) lack credibility for farmers. For example, farmers may take the land-tenure situation into account when deciding where to locate plots for a variety or fertiliser trial, in order to observe the influence of this factor.

- **Systematic evaluation.** It should be clear to everyone why and how a certain conclusion was reached. Farmers should be able to explain to other farmers the criteria they used and on what information they based their conclusions.

IMPROVING FARMERS' EXPERIMENTAL DESIGN

Farmers' own ways of experimentation may have important limitations (see Unit 3.A, pp. 145–55). For example, a farmer in Mali compared two varieties of millet, one on a plot previously planted with millet, the other on a plot previously planted with beans. When the latter gave higher yields, the premature conclusion was drawn that this variety was superior (Stolzenbach, 1992). Box 3.4 lists some important issues of experimental design that can be considered together with farmers.

According to Bunch (1989), experimentation should be on a limited scale for the following reasons:

- it reduces the level of risk;

> **BOX 3-4 DESIGN ISSUES IN IMPROVING FARMERS' EXPERIMENTATION IN CROPPING**
>
> - **Suitable location**
>   - similar to farm situation
>   - uniform situation, e.g. soil
>   - protected from theft, animals, other disturbances
>
> - **Limited experimental plot size, depending on:**
>   - type of crop involved, e.g. tree seedlings in bags, chillies on beds, cowpeas on 10 x 10 m plots
>   - type of experiment, e.g. varietal trials plots of 10 x 10 m, but soil cultivation trials 100 x 20 m
>
> - **Good demarcation and separation of plots**
>
> - **Elimination of border effects**
>   - for example, when harvesting the experimental plot, exclude the outermost 0.5 m
>
> - **Several replications**
>   - doing a trial with a new idea on five farms rather than on one farm gives a better basis for drawing conclusions, if farm conditions are similar
>
> - **Allowing for a control**
>   - answer the question: "With what do we want to compare the experiment?"
>
> - **Limiting the experiment to only one issue or variable**
>   - answer the question: "What are the differences between the plots or treatments that we are studying?" (preferably, there should be only one difference)
>
> - **Systematic monitoring**
>   - answer the question: "What information do we have to collect to be able to draw conclusions from the experiment?" (compare Unit 3.D, pp. 168–80)
>
> *Source: adapted from Beingolea (1994) and PMHE (1993)*

- it enables a farmer to carry out several experiments simultaneously, thus providing more opportunities to learn;
- the rest of the farmer's land can serve as a natural control plot;
- it is also easier for poorer farmers to participate in testing a new technology.

Werner (1993) points out that, while relatively small plots (30–50 m^2) may be used for testing a new variety or a fertiliser, larger plots (50–100 m^2) are needed to test new cropping patterns, especially intercropping, or soil fertility practices. Sometimes, bigger plots may be needed to allow farmers to observe and measure results using local means. For example, planting a new maize variety may give an increase in yield of 20%, or an increase from 4.0 to 4.8 kg in a test plot of 20 m^2; this amount could hardly be recognised without the help of a balance. In a test plot of 100 m^2, the difference in yield (4 kg) could be measured more easily using local means.

Especially in smallholder agriculture, it will often be difficult to make several replications on only one farm. For this reason, replications within fields (as usually applied in researcher-managed trials) are normally avoided. Instead, "replications" are made across farms which are spatially clustered so that participating farmers can easily visit each other to observe and discuss their experiments. In the initial phase, when fieldworkers are learning to work in a participatory mode with farmers, the number of experimenting farmers per fieldworker should be limited (e.g. a maximum of 2–3 villages with 5–8 experimenting farmers each).

The guidelines given in Box 3.4 will lead to experiments with only one factor or variable, which is how farmers themselves often work (Unit 3.A, pp. 145–55). However, many farmers are also managing and evaluating complex trials involving numerous variables and/or comparing 5–10 varieties at a time.

FARMER EXPERIMENTAL DESIGN WORKSHOPS

Those farmers who show an interest in trying out one or more of the technical options identified during the participatory analysis of their situation are brought together to discuss how they could design and implement the experiments. Activities during such an experimental design workshop may include:

A simple trial layout

- discussing one or more *case histories* of experiments done by members of the group (or by farmers visited elsewhere), discussing how they did the experiment and why they did it that way. This may involve spontaneous reports about experiments being given during the meeting, or prepared cases being presented by farmers who were visited previously by a PTD facilitator, to explore how the farmers did their experiments (see Unit 3.A, pp. 145–55). A slide series or video may support such a presentation;

- systematically going through a *series of "prompting questions"* to stimulate farmers to discuss and decide upon the design, implementation and monitoring of the specific experiment they have in mind. It can be fruitful to invite one or more farmers from other areas with experience in this type of experiment to participate in the workshop (Simaraks *et al.*, 1986);

- providing, if still required, additional insight into experimental design. World Neighbors reports that discussions with Bolivian farmers on small-scale experimentation led them to ask for a special training on experimental design. A simple training manual was then developed (Ruddell & Beingolea, 1995).

In discussing the implementation of experiments, some additional important issues to be considered are:

- **Choice of farmer-experimenters:** Should all interested persons be involved in the first trials, or only a few? And at a later stage? If only a few, according to what criteria should they be chosen? Consider also criteria such as gender and economic position;

- **Organisation and timing of the experiment:** What would be the best time to start the experiment? Do all farmer-experimenters start at the same time? What inputs are needed, and where can they be obtained?

- **Monitoring and evaluation:** What information do we need to collect? How do we collect this? Who will do what, and when?

It is important to include monitoring and evaluation in the discussions at this stage. Difficulty in answering the above questions indicates that the experimental design has not been thought through sufficiently. Unit 3.D (pp. 176–88) on monitoring, analysis and evaluation provides further guidelines on designing an effective recording system.

GOING DEEPER INTO TRIAL FORMATS

Here, some simple trial formats are briefly described in order to create a basis for discussion and reflection with those participants who have not done systematic experimentation before. By no means is it intended that designs for farmers' experiments should be restricted to the formats presented here (see also Werner, 1993):

- **Individual application of newly acquired materials:** to test a new piece of equipment, a new crop or a new animal breed, farmers may decide to obtain the equipment, seed or breed and let some of their group try it out in whatever way they consider best. At certain intervals, they come together to observe and discuss how the experimenting farmers are using the new materials and the results obtained. In this case, both experimental and non-experimental variables will vary greatly among the participating farmers.

- **Paired comparison:** farmers make a planned comparison between two possibilities, e.g. they compare two plots marked prior to the onset of the season which are cultivated with and without manure application or with a traditional and a new variety, or they compare two groups of animals given different types of feed. This format normally requires that the groups of animals are similar in age and weight and are kept under similar husbandry conditions, or that the same crop and variety is used for all paired comparisons, with the same seeding rates across farms and the same planting date for each plot in the pair.

- **Before and after the experiment:** rather than comparing two plots or groups of animals (with and without treatment), a situation can also be compared before and after treatment (e.g. soil erosion by run-off before and after planting grass strips along contour lines). In this case, information on the same variables must be collected in both the "before" ("baseline") and the "after" situation. The experiment could entail one factor, or a combination of innovations of which the total impact will be evaluated.

- **Expanded paired comparison:** in principle, a paired comparison could also be made between two different combinations of practices; for example, between a plot with traditional practices and a plot with a combination of "improved" practices. Also the number of treatments in each comparison can be expanded, where necessary and appropriate; for example, to compare several varieties. In both cases, but especially the first, it is important to specify the differences between the two sets being compared. The assistance of a statistician may be needed to determine how to evaluate the results.

- **Superimposed treatment:** the difference between this format and previous ones is that farmers sow their fields as usual and the experimental variable is later "superimposed" in selected parts of the field, e.g. an extra weeding, a top dressing. The treatment plots are compared with one or more adjacent plots of the same size. Such a design may be useful as a response to research opportunities that arise during the growing season. For example, in a meeting to discuss a maize variety experiment, farmers may express a desire to check the effect of weeding. They can then decide to leave part of their standing crops unweeded and mark this part of the field.

- **Stepwise or "add-on" design:** in this case, one trial includes various innovations, each being one step on the way to a full technical package. The trial is organised in such a way that farmers can evaluate the effect of each "addition" to the package separately. The innovations are added on in the order of the level of expected returns from that component.

FOUR WARNINGS

There is a danger that PTD facilitators, especially those with conventional agronomic training, get "carried away" with all the possibilities of improving farmers' experimental design. This can be prevented if one realises that the design chosen should:

- be the *farmer's design*. S/he should "own" the design and be responsible for managing and implementing the experiment. This also implies that gifts or payment of time invested are inappropriate, and that the experimenting farmers are responsible for acquiring the required inputs. If farmers continue to ask for large amounts of free inputs, this is a clear signal that they do not yet consider the programme to be their own;

- be *simple* and easy for the farmers to understand and manage. The experiment should address the major factor first. Low priority is given to manipulating variables that require a major change in the

Farmers own the experiments and carry the risks involved

level of management and (external) inputs required. When several variables are of high importance, stepwise testing (see below) should be considered;

- be *flexible* and allow for later adaptation. What might seem to be an appropriate design could need adaptation later, for reasons related either to the experimentation or to individual farmers' conditions and preferences. Monitoring and improving the experimental methods is an important part of the learning process. Deliberate experimentation with and adaptation of trial designs should be encouraged. The possibility that some farmers will drop out should be anticipated. The initial number of experimenters per village should not be too small (minimum 5), and possible reasons for dropping out should be discussed with the experimenters;

- lead to clearly *visible and significant results* according to local standards. This means, for example, that the trial plots should be neither too small (differences cannot be observed properly) nor too big (too risky). In some cases, results cannot be observed easily and have to be made visible, for example, by calculating the costs and benefits of each treatment together with the farmers.

STEPWISE TESTING TO DEVELOP FARM SYSTEMS

In most cases, tackling problems requires multiple innovations that are interrelated and, together, form a solution to the problem and/or create a significant and sustainable change in the farm system. Conventional on-farm research was therefore normally concerned with a newly developed technology, such as a new crop variety, and the package of recommendations that accompanied it (e.g. plant densities, application of fertiliser and pesticides). On-farm testing was primarily meant to find out how farmers reacted to the package and to identify constraints to adopting the package.

In PTD, the starting point is the farmer with his/her multiple aims and the complex and site-specific farm system. It is not assumed that complete packages, to say nothing of entire cropping and livestock-keeping systems, can be designed fruitfully by outsiders. The farmers themselves hold the keys for developing, evaluating and validating these systems.

PTD is not restricted to simply comparing a few technical options. The relationship between innovation and the development of the entire farm system should be kept constantly in mind and form an integral part of designing and evaluating the trials. Outsiders and farmers will assess the contribution of each innovation to the solution of the larger problem, and how it relates to enhancing the sustainability of the farm system as a whole (Unit 1.B, pp. 30–40). Farmers are given the opportunity to develop new insights into the relations between the various technologies, the various components of their farm and the underlying biological processes. This creates new avenues for sustainable development of their farm within the larger agro-ecological system.

In this perspective, farmers' experimentation takes on the character of a chain of small tests and improvements, each building on the previous ones. At one moment in time, a farmer may be experimenting with a number of incremental and seemingly unrelated changes in his or her farm plan and practices and may incorporate only a few of them. However, after some years, this may result in major shifts in the farm system and management. In on-farm research, the "stepwise" or "add-on" trial design is sometimes used: various innovations are included in one trial, each innovation being one step on the way to a full package. In PTD, the "added-on" variables will probably be tested by the farmer in later trials (if the trial with the first variable was promising), or this will be done simultaneously by other farmers who consider one of the other variables to be more important.

LEARNING ACTIVITY 1

PROMOTION OF FARMERS' EXPERIMENTATION: A CASE STUDY

2½ hrs

Objectives:
- To enhance awareness of different strategies that can be employed in supporting farmers' experimentation and of the importance of flexibility in using them.
- To understand the pros and cons of aiming at improving farmers' experimental designs.
- To gain an initial understanding of issues in systematic experimentation.

Setting/approach:
- Workshop setting in which small groups discuss the case study and report back results for discussion in the plenary session.

Materials:
- Pens and paper for the small groups; newsprint sheets with markers for reporting to the plenary; optional handout of a case study such as that given in Box 3.5.

Procedure:
- After explaining the objectives of this learning activity, present a case in which farmers' experimentation was systematically encouraged. This could be an oral presentation, using audiovisual materials, or in a written form. It is better to select a case study from one of the participants or from the local area. Alternatively, the Sri Lankan case of Box 3.5 may be used. If a local case is used, care should be taken to present it in such a way that relevant issues for discussion are evident.

- Participants form small groups of 3–5 people to study the case and answer assigned questions, such as:
  - What did the staff do to encourage experimentation by the farmers?
  - What good points do you see in their approach?
  - In which areas (if any) could the methodology and experimental design be improved? Suggest concrete improvements.
  - Could you apply a similar approach in your own work? If not, why not?

- Groups report back to the plenary session, presenting the main points on newsprint sheets. All groups may be asked to give their response to the first question, followed by discussion, then to the second question, followed by discussion, and so on.

- In the plenary discussion, several crucial themes should be raised:
  - whether there is a need for "proper" experimental design;
  - usefulness of the recording of data by farmers – for their own use or to help researchers? In whose interest is the experiment done?

BOX 3-5 CASE STUDY OF FARMERS' EXPERIMENTS ON GERMINATION OF TEAK SEEDS

The problem:
Teak is a species that most farmers in the Dry Zone of Sri Lanka want for their homegardens, since they value its timber both for home use and sale. Establishing trees usually involves some cost, either for purchase or to travel to teak-producing areas. The alternative of growing from seed runs up against the problems that teak seed is difficult to germinate and that few farmers are familiar with doing this.

Objectives:
The activity described below was intended to answer the question: "What is/are the best way(s) to germinate teak seed for small-scale nursery production?".

The programme also had some non-research objectives: to increase the farmers' confidence and experience in raising tree seedlings to meet their own needs; to stimulate interest in experimenting in groups, possibly as the start of a wider programme of experimentation; and to grow some teak.

Methodology:
Having established through informal farm visits that farmers were interested in trying to raise teak, two agroforesters discussed in group meetings the problems of germinating teak seeds. In most groups, a few farmers had heard of one or two techniques to enhance seed germination. Very few had ever done it in practice, because they were not aware of the technical details and lacked confidence in their own ability to use the techniques. The agroforesters pooled farmers' ideas and their own, and identified three methods of germinating teak, and their advantages and disadvantages:

1. *The traditional burning method.* Teak fruits are placed on a shallow bed of paddy husk or other burnable material, covered with the same material, and set alight. The aim is to burn the hard outer seed coat without damaging the seeds inside. After burning, the fruits are planted. This method is known to many farmers, even if they have never tried it.

2. *Soaking/drying method,* as used by the Forest Department. Fruits are alternately soaked and dried over a period of two weeks before planting. Most farmers were keen on this method, because it seemed likely to be reliable.

3. *Opening the fruit to expose the seed.* This was unfamiliar to farmers and was demonstrated by the agroforesters, using a sharp knife.

Farmers were given complete freedom to choose how much seed they wanted and which method(s) to use. At this point, the idea of experimenting arose spontaneously in several groups — with either whole groups dividing up the methods between them, or individuals volunteering to try one or more methods. Farmers were then requested, in return for the seed provided, to keep records of the methods used and germination results. To assist them, the agroforesters prepared and distributed record sheets. As further motivation for keeping records, the agroforesters explained that these would help to decide on the best method and would help pool the information from all the groups, so that the results could benefit a large number of people.

What happened:
All farmers who received seed tried out one or more treatments. The agroforesters made repeated individual farm visits to motivate, help with problems and record what was being done. These notes helped in cases where farmers did not keep their own records, or kept them inaccurately. In particular, it was found that people's memories of dates were unreliable; if the recording sheets were not filled out daily, the data were not very reliable. In many cases, the farmers did things differently from how they had said they would: some who had volunteered to try several methods in fact tried only one, whereas others who had not appeared interested in experimenting tried several methods. In a few cases, farmers asked for, and were given, more seed for further experimentation.

The results were analysed both within and between the groups of farmers. This involved looking at the yield of the different methods, i.e. how many seedlings were obtained from a certain number of seeds, as well as the convenience and risk of the different methods and how this affected yields. To be able to draw any conclusions, farmers were assisted in doing a simple matrix scoring of the three treatment methods on the basis of the above criteria. In a few groups, results were not conclusive and the decision was therefore taken to continue experimenting, with improvements in the experimental design.

Source: adapted from PMHE (1993),
based on fieldwork by Stephen Connelly and Nicky Wilson

LEARNING ACTIVITY 2
THE MINIMUM-REQUIREMENTS QUESTION

2½ hrs

Objectives:
- To increase awareness of different options in strengthening farmers' experimentation.
- To stimulate critical reflection on preconceptions about "proper" experimentation.

Setting/approach:
- Groupwork around the question of the extent to which farmers' experimentation should be made more scientific.

Materials:
- Handout with a list of requirements for experimentation to meet (scientific) standards (e.g. prepared on the basis of Box 3.4). Newsprint sheets and markers for reporting results of groupwork.

Procedure:
- The objectives are explained, for example, in terms of the need to clarify the debate on "making scientists out of farmers".

- Participants form small groups to study a list of requirements for "proper" experimentation and to answer the following questions:
  - (for each issue on the list) Is this is really a basic requirement which should be incorporated into farmers' experimentation to ensure valid results?
  - Why is (or isn't) this a basic requirement?
  - What modifications or additions can be made to the list?

- Groups report and discuss their findings in the plenary session, possibly by reporting and discussing one issue from the list at a time.

- In the plenary discussion, several key issues will arise:
  - Is there a need for quantitative analysis?
  - What are valuable results? To whom are they valuable: the experimenting farmers themselves, other farmers who may wish to learn from the experiments, or the researchers involved?
  - Is there a possibility to combine farmers' and researchers' agendas and interests?

LEARNING ACTIVITY 3

THE OLD LADY WHO ONCE HAD A PAIN:
A Story Highlighting the Need for Systematic Experimentation

30–60 min

Objectives:
- To make participants realise that you learn most from an experiment if you study only one innovation at a time.

Setting/approach:
- In a workshop session a simple story is studied and discussed.

Materials:
- The story of "The lady who once had a pain", from the comic strip on p.166.

Procedure:
- The story is presented in a creative way.

- Participants are challenged to reflect on the story in two steps. First, they look at the actual case. Questions for reflection may include:
  - Why was the old lady unable to respond to the old man's question?
  - Which of the cures worked?
  - How can we know? Why?

- In the discussion it may be concluded that, because several cures were tried at the same time, we cannot tell whether the old lady was cured by one alone, by a combination of two of them, or by all three jointly. In order to be sure, one would have to test each cure separately on three people with the same pain, or on one person with the same pain at three different times.

- The second step is for the participants to "translate" the results of this reflection to their work in experimenting with agricultural innovations:
  - What happens when we try out several innovations on the same plot at the same time?
  - If we do well, or even if we do not, how can we determine what caused it?

  Both reflections usually take place in a plenary discussion but, alternatively, can be prepared in some form of small-group work.

- In the final discussion one may conclude that a farmer who tries out a new fertiliser along with some new seed planted using a new spacing in his plot will not be able to compare the results with the normal plot. One must always compare one innovation at a time. By preparing several small plots, it is possible to compare the outcome of each innovation separately in relation to the normal plot.

Variation:
- The story can easily be turned into a play. Of course, the details can be adapted to fit local practices.

Source: PTD Circular 4 (1995); adapted from SIMAS (1995)

THE OLD LADY WHO ONCE HAD A PAIN – LEARNING ACTIVITY 3 (continued)

LEARNING ACTIVITY 4
DEVELOPING GUIDELINES FOR DESIGNING EXPERIMENTS WITH FARMERS

2 hrs

Objectives:
- To develop sensitivity to the weak and the strong points in certain experimental designs and to think in terms of "balanced" rather than "correct" or "false" designs.
- To learn guidelines for systematic experimentation by farmers.

Setting/approach:
- Debate in a workshop setting.

Materials:
- Handout with cases of farmer experiments; examples are given in Box 3.6, but the exercise is more effective if cases are prepared on the basis of local experience.

Procedure:
- Participants are given a handout with descriptions of simple experiments.
- Participants form two groups: one is asked to identify the strong points of each case, the other the weak points.
- After the groups have prepared themselves, representatives of both groups discuss in the plenary session the weak and strong points of that case. Once the representatives run out of arguments, other group members may raise other issues. Each discussion is followed by jointly defining one or two learning points concerning the basic principles of experimentation.
- The findings are summarised and integrated, preferably in a visual form.

Variations:
- First give the participants a list of key issues and questions about systematising farmers' experimental methods, then ask them to analyse the cases with the help of this list.
- If time is short, the cases may be studied individually. This stimulates all participants to define their opinions.

BOX 3-6 CASES OF FARMERS' EXPERIMENTATION

A small-scale farmer in western Zambia wants to try out a more drought-resistant variety of maize and asks for seed to sow an entire field (1 acre) with the new variety. *(Issues for discussion: pros and cons of whole field vs small trial plot; use of inputs that are normally unavailable to the farmer; whether outsiders should provide essential inputs for the trial and/or share the risks of experimentation; how to compare alternative options.)*

A woman farmer in southwestern Nigeria plants a mixture of cassava and an improved maize variety (less susceptible to a certain pest) in a small trial plot (50 m²) in the centre of a plot with a mixture of cassava and traditional maize. Because of drought, she does not tend the field, nor does she harvest the maize crop. *(Issues for discussion: what to take into account when locating a trial plot in a field; whether the absence of good management of the plot is, in this case, a desirable feature of the trial or something to be avoided; how monitoring and assessment of the trial is influenced by drought, and what can be done about it.)*

A group of farmers in Peru want to compare three new cassava varieties with their local one. They make a trial plot in the middle of their field consisting of 8 sections of 5 x 10 paces, as they expect that different fertility conditions will give different yields and root quality; they compare two levels of fertiliser application on their local cassava variety and the three new ones. *(Issues for discussion: complexity of farmers' experiments vs ability of farmers and outsiders to lay out the trial without major mistakes and to distinguish between all treatments during recording and assessment; what can be done about this complexity; farmers' holistic vs scientists' reductionistic approach to evaluating and dealing with complexity; need for statistical analysis.)*

In Sri Lanka, Rani is interested in growing chillies and is always looking for new things to try out. During earlier work as a labourer on a research farm, she was told that chilli seeds sown on beds produce strong plants. She decides to try this out by sowing one third of her seed on a well-prepared bed and the rest in her usual way. She looks after both in the same way. When the results in the seedbed turn out to be no better than usual, she decides that making seedbeds is not worth all the work. *(Issues for discussion: the value of 1-season trials, replications, what data to collect, management of experimental plot similar to that of other plots.)*

Also in Sri Lanka, Abekoon has kept chickens (layers) for the last three years. As the price of feed recently rose dramatically, he is seeking cheaper alternatives. He makes a smaller, separate pen for 10 of his 100 layers in order to try a new, locally-made feed. He gives the 10 layers a mixture of the earlier feed and the new one; the other 90 are given only the earlier feed. He takes good care of all the birds. After three months, he finds that the 10 "experimental" layers produce as many eggs as the others, so he decides to give all 100 chickens the new mixture. *(Issues for discussion: possibilities and risks in experimentation with livestock, what is the treatment: the new feed or the mixture prepared by Abekoon?, length of the experimental period.)*

LEARNING ACTIVITY 5
SIMULATION: "FARMER DESIGN WORKSHOP"
3 hrs

Objectives:
- To familiarise participants with the main issues and processes involved in a design workshop.
- To develop skills in the use of "prompting" questions to stimulate farmers' discussion and decision-making about experimental design.

Setting/approach:
- Role play in a workshop setting.

Materials:
- Handout with guidelines for implementing a design workshop, which can be prepared on the basis of the relevant pages of the discussion section of this unit.

Procedure:
- In small groups, participants discuss a handout which offers guidelines for implementing a design workshop and describes the main items to be discussed during the workshop (30 minutes).
- Some participants are selected to act as facilitators while the others act as interested farmers. The "facilitators" are given about 15 minutes to prepare themselves. At the same time, the "farmers" are asked to think of the attitudes and skills of farmers in the area, and to decide what technical options they are interested in experimenting with. The trainer pays special attention to this preparation by the "farmers" in order to ensure a realistic simulation.
- Two or three participants may be asked to make observations concerning, for example, the process of the meeting, the role of the facilitator, the use of the questions. A simple observation sheet may be developed and given out to the observers.
- The design workshop is held (about 30–60 minutes is enough for the learning purpose).
- Experiences and observations are shared. The final discussion is focused on the skills:
  - dos and don'ts in moderating such a design meeting;
  - further development of an adequate set of prompting questions and proper phrasing of the questions.

Variation:
- A similar workshop can be held directly with farmers. Generally, however, a classroom session should precede the meeting with farmers, to give fieldworkers the opportunity to practise some of the required skills.

LEARNING ACTIVITY 6
MUTUAL CONSULTATION ABOUT SELF-DESIGNED EXPERIMENTS
3 hrs

Objectives:
- To enhance participants' understanding of the main issues involved in designing an experiment.
- To strengthen their skills to assist others in this respect.

Setting/approach:
- Groupwork and role play in a workshop setting.

Materials:
- Handouts with main methodological issues in designing simple experiments (optional); newsprint sheets and markers.

Procedure:
- The main principles and methodological considerations for designing experiments with farmers are introduced and discussed; also some examples of trial formats. A handout (on the basis of Box 3.4) may be distributed.
- In small groups of 3–4, the participants design a simple experiment, taking a local farm situation and the objective of the trial as defined by a group member as the point of departure. As the exercise is more interesting if the groups choose different experiments, the trainer could monitor this.
- In turn, each group presents its design while one other group acts as "facilitator" and asks questions about the reasoning behind it and problems encountered. The groups may also assist in improving the design of the trial.
- After each presentation and discussion, participants reflect briefly on how the "facilitators" assisted the "farmers" in improving their designs.
- The main learning points from the mutual consultations are summarised by the trainer.

LEARNING ACTIVITY 7
RECONSTRUCTING FARM DEVELOPMENT

4 hrs

Objectives:
- To enhance understanding of the process of gradually developing a more sustainable farm system over time.
- To develop skills in the stepwise planning of experiments with farmers.

Setting/approach:
- Study of case histories by small groups in a workshop setting; in the variation, possibly also on the farms of selected farmers.

Materials:
- One or more case histories of farmers who have gradually transformed their farms into integrated ecological farms. Ideally, the cases should be developed from local experience, but a simple example is given in Box 3.7 below.

Procedure:
- Some basic principles of ecological farming are briefly reviewed. If the participants are not familiar with them, these principles must be discussed in detail (Unit 1.B, p. 32).
- In small groups, the participants study case histories of farmers who have gradually been transforming their farms into more integrated ecological systems. The participants trace the succession of experiments and innovations that have been made by these farmers over a number of years. They may also be asked to suggest a possible next step in the chain of innovations.
- Results of the groups are reviewed briefly in a plenary session.
- Participants are then asked to think through a sequence of experimentation and innovation for the most common farm type in their working area, starting from the problems and technical options in which the farmers are most interested.
- Outcomes of these individual deliberations are presented and discussed, preferably with the participation of some farmers from the area involved.

Variation:
- Identify local ecological farmers and ask participants to reconstruct, together with these farmers, their farm development (either on the farm or in the workshop).

BOX 3.7 MR CASAS'S FARM: AN EXAMPLE OF GRADUAL TRANSITION TO ECOLOGICAL FARMING

On about three-quarters of his one-hectare farm, Mr Casas, a farmer on the Philippine island of Mindanao, had been growing wetland rice using all the recommended inputs. After attending a course on organic agriculture, he decided to try to reduce his dependency on chemical inputs, most of which he had to obtain on credit.

In the first year after the course, he started by planting the cuttings of Madre de Kakaw (*Gliricidia sepium*) which he had obtained during the course, as he wanted to try the leaves as an insect repellent against caseworm.

The next season, he made a more drastic change in his practices by not burning his rice straw but instead bringing it all back on the fields, letting it rot there and working it into the soil during the next land preparation.

After harvesting the subsequent rice crop, he decided not to leave the newly-harvested fields fallow but instead to broadcast 6 kg of mungbean seed, mainly to improve soil fertility.

Being confident of the impact of all these measures, he decided to reduce nitrogen (N) fertiliser in the next year's rice crop to only 25% of the normal dose. This leaves him with the further challenge of also eliminating the remaining part of N fertilisation, as well as of some chemical poisons for rat control and some herbicides.

Source: Escazo & Olbers (pers. comm.)

UNIT 3·C / FARMER-EXPERIMENTER GROUPS

OVERVIEW OF THIS UNIT

EXPECTED RESULTS
After completing the learning activities in this unit, participants are expected:

- to be aware of the role that farmer groups can play in strengthening farmer experimentation;
- to have developed a balanced view on the individual versus the group approach, and on strengthening farmers' experimentation;
- to be familiar with the most important principles in promoting farmer-experimenter groups.

MAIN CONCEPTS
- **Farmer-experimenter groups:** Collaboration and mutual exchange of experiences among farmer-experimenters.
- **Self-selection, gradual organisational development and self-organisation** as basic principles in promoting farmers' groups.

TRAINING METHODOLOGY
The participants are encouraged to analyse in detail and actively discuss various aspects of promoting farmer-experimenter groups.

LEARNING ACTIVITIES
1. Writing guidelines for group development
2. Debate on the role of farmer groups: a game of opposites

DISCUSSION

THE ROLE OF FARMER-EXPERIMENTER GROUPS IN PTD
PTD focuses on transforming local experimentation from being relatively *ad hoc*, unorganised and individual to being more focused, systematic and organised into a community process of technology development. Thus, PTD is a process not only of producing appropriate technologies but also of building self-help institutions for agricultural development. Although such institutions can take different forms (see also Unit 4.B, "Sustaining the Process"), farmer-experimenter groups very often form an important part of them.

The term "farmer-experimenter groups" refers to groups of farmers who meet periodically to discuss farming problems and potentials, select technologies for testing, and discuss and assess the results. Group members also play an important role in informing other farmers about their results and the experimental methods used. Outsiders may be invited to participate in some of the group activities, but the group is self-governed.

In the PTD process, the experimenter groups also contribute to:

- deepening the situation analysis and problem identification through the confrontation of opinions;
- developing a joint understanding of main constraints and opportunities;
- enhancing farmers' own experimental work through exchange of results, replications across farms, and increasing the range of technologies that can be tested;
- linking with support institutions (governmental and non-governmental) to obtain their services; and
- ultimately, influencing government policies.

Of course, unless farmers find significant value in jointly planning and evaluating experiments, they will prefer to spend their time doing something else. Group cooperation must lead to results which they could not have attained as individuals. For a detailed discussion of the role of farmer-experimenter groups, see Heinrich (1993).

A farmer-experimenter group is often mainly a platform for discussion and sharing information, while each participant decides whether to implement his or her own trial. In other cases, the group may also take on other functions, such as organising the acquisition of inputs (e.g. local seed production) or helping each other to overcome certain labour or cash constraints related to the technology tested, i.e. the group becomes more oriented to general agricultural development.

In most communities, local institutions for agricultural development activities will exist. Men and/or women farmers may, for example, take turns in working together on each other's fields. Some of these local institutions may have a specific role in experimental work (Brouwers, 1993). Farmer-experimenter groups will, where possible, build on these.

FORMATION OF FARMER-EXPERIMENTER GROUPS
Groups can be formed in many ways, depending on the local culture, the farmers' interests, and the working approach of the supporting agency. In general, however, farmer-experimenter groups emerge slowly during the initial diagnostic and planning stages. The outsider helps to make sure that all categories of the population – especially the poorer people – can take an active part in analysing the situation and looking for things to try out (compare Units 1.E: pp. 59–66, 2.B: pp. 102–12 and 2.C: pp. 113–28). The facilitator encourages those farmers who wish to try out new options to collaborate in designing and organising the trials. Three principles guide the facilitator in such efforts – those of self-selection, gradual organisational development, and self-organisation.

Self-selection implies that some of the farmers who initially took part may leave the group when it becomes clear that involvement does not lead to the easy gains they expected, or when they see who else is participating. On the other hand, others may join when they hear that, among the priorities selected for the trials, there are subjects of interest to them, or when they hear that certain other people will participate. This process results in groups of farmers who share a *common interest in experimenting with certain technical options*, be it in testing a specific technique related to a particular crop or problem (specific focus), or in testing various options of different natures (broad focus). The composition of the groups and their degree of homogeneity will vary. In Honduras, a group of farmers experimenting with cassava consisted only of men, whereas a similar group in Zaire consisted only of women, because of differences in the gender-specific division of labour (Box, 1989). The more homogeneous the group in terms of self-defined interests and perceived problems, the more effective the group process is likely to be.

Gradual organisational development is important in preventing the imposition of outsiders' organisational models. It implies that the size of the farmer-experimenter groups may grow and that their interests may diversify over time. As an example of this, Table 3.1 summarises the group development stages experienced in the project area of the PMHE project in Sri Lanka. This may create a need for new, smaller, more focused interest groups. Experience shows that small and fairly homogeneous groups (in terms of main interest, gender, kinship, level of resources) are usually the most successful. Homogeneity is especially important when the group takes on functions beyond the joint planning and discussion of experiments (Verhagen, 1984; Gubbels, 1988). Groups may also undergo a development from rather informal get-togethers to a formalised organisation. The international research institute CIAT, in Columbia, encourages the emergence of more formal farmer-researcher groups (Ashby *et al.*, forthcoming). Among its reasons for doing so is the need to strengthen the capacity of such institutionalised groups to receive and manage government research funds. Farmer-experimenter groups may have overlapping memberships and develop mechanisms of information exchange at the community level, and/or gradually link up with other types of farmer organisations at the local and regional levels.

Self-organisation is a principle based on the premise that farmers themselves are in the best position to organise their cooperation in the most effective way. They know their own situation and can take into account the particular circumstances of everyone who is involved, seasonal variations in time availability, and cultural factors which affect group activities. Experimental groups are therefore likely to be more successful and sustainable if, from the very beginning, they set their own rules and develop in their own way and at their own speed.

CHARACTERISTICS OF SUCCESSFUL SELF-ORGANISATION

Some characteristics of successful group formation and group activities in agricultural experimentation are:

- **Common interest and focus:** the group consists only of persons who normally perform the activity to which the technology is related, who already have relevant knowledge based on traditional practice, and who have in common a strong personal interest in trying out certain technologies;

- **Self-selected coordinator:** the participants select their own group coordinator and define clearly what is expected of him or her during a defined period, e.g. the coming year;

- **Periodical meetings in the test fields:** the group works out a schedule of meetings, based on exchange visits to each other's farms. All members show and discuss their experimental activities on an equal basis. The rhythm and number of meetings depends on the type of experiment and the "critical" moments in its development: times when certain issues can best be observed and discussed;

- **Self-organisation of the joint activities:** the group itself organises the meetings, sets the dates, invites outsiders and handles related logistical and financial issues;

- **Well-prepared meetings:** the group defines for each meeting what should be observed, measured, discussed and done;

- **Documentation and sharing:** the group records in some way the results of each meeting and informs other people in the community of these results;

- **Periodical self-evaluation:** the group periodically (for example, at the end of each season) evaluates how it is functioning and what it has achieved, and adjusts its objectives and operational procedures accordingly.

TABLE 3·1
GROUP DEVELOPMENT PROCESS: EXPERIENCES FROM SRI LANKA

STAGE	ACTIVITIES	GROUP CHARACTER	INPUT PROJECT	TIME
0	Start with PRA activities to learn situation; hold problem discussions in general, also solution discussions; motivate to work together in group (mostly neighbourhood groups?).	Different forms of informal group discussions.	Various staff involved in PRA activities.	2–4 months
1 Pre-group	Interested people come together and go through a series of meetings (3–6). The project's and group members' objectives are clarified; more detailed discussions of problems and possible solutions; prioritise solutions.	Very informal.	Intensive guidance through social mobilisers	1–2 months
2 Emerging group	Start of first activities, e.g. labour sharing, group fund, homegarden development experiment, trying of crops other than paddy in rice fields. While doing this, attitudinal changes, learning of recording skills, unity and trust develops, leadership emerges. Criteria for effective initial activities include: ■ use own present resources ■ do not depend on capital from outside ■ use known or easily learnt technologies – "simple" ■ do not require strong unity ■ mostly encourage some kind of collaboration.	Informal to semi-formal. Name is chosen, certain rules develop, minutes of meetings are noted. In some cases the positions of chairperson, secretary and treasurer rotate each week; in other cases, there are three more permanent positions.	Frequent attendance at meetings, support moderation of meetings, various training-type inputs on various activities (record-keeping etc.).	2–3 months
3 Developing group	The group forms a constitution to better organise and formalise its functioning. More complicated activities, requiring more capital and stronger unity, can be taken up: e.g. credit to members, livestock, common income-generating activities, common farm, more experimentation, sharing experiences with other units/blocks.	Semi-formal to formal. In some cases, continued rotation of chairperson position.	Less frequent attendance at routine meetings; technical support, training in selected activities. General guidance in learning process.	6–8 months
4 Effective group	Groups on their own take up more complicated activities such as joint marketing, common nurseries.	Groups function mostly on their own, only subject-specific input from project. Several groups may coordinate activities to be more effective, and discuss developing towards federation of groups.	Subject-specific support.	?

Source: PMHE (1994)

PLANNING GROUP MEETINGS

Group meetings have to be planned by the participants in advance. Issues that need to be considered when planning group meetings and exchange visits include:

- Who should participate in them: just the farmer-experimenters? also farmer-experimenters from other villages who can contribute their observations and experiences? interested farmers from the community or elsewhere who want to be informed and to learn? other relevant outsiders who could contribute expert observations and comments (e.g. researcher, middleman, mill owner, extension worker)? local leaders?

- What can/should be measured, observed and discussed during this meeting, and how will we do this? Who will guide the meeting, and who will write down the results for later use?

- Where do we meet? Do we visit all experiments, a selection of them, or only one?

- How do we organise this in practical terms: time schedule, transport needed, food/refreshments, materials needed; who will do what?

When the people attending a meeting differ greatly in their interests (and consequently in their assessment criteria) or when the group is too large, it is sometimes better to form subgroups to give everyone a chance to express his or her observations, and report later to the whole group. If the number of questions to be looked into is large, different groups may choose to focus on different questions.

The outsider initially assists in moderating these group meetings, especially in stimulating farmer-to-farmer interaction. S/he may also play a role in recording farmers' observations and helping to structure the outcome of the discussion.

LEARNING ACTIVITY 1
WRITING GUIDELINES FOR GROUP DEVELOPMENT

2½–3 hrs

Objectives:
- To enhance participants' awareness of the important factors influencing successful cooperation among farmers in their working area.
- To develop simple guidelines for facilitating development of farmer-experimenter groups.

Setting/approach:
- Workshop setting with considerable time spent in small groups.

Materials:
- Newsprint sheets and markers to present main findings of groupwork for the plenary discussion.

Procedure:
- Participants are divided into small groups, preferably according to different socio-cultural situations. Each group is asked to reflect on the history of cooperation among farmers in their area and to analyse real cases of success and failure they know of. This will lead to the questions: What are the important factors that favoured successful cooperation? On the basis of this analysis, the groups draw up recommendations, e.g. in the form of dos and don'ts, for facilitating group formation and development in this particular situation.

- The subgroups present their recommendations to the others, acting as if these were fieldworkers new to the area, who have to be acquainted with the socio-cultural situation and the organisational patterns and processes.

- The differences between the subgroups are discussed. Participants are asked to put themselves in the shoes of farmers as much as possible. If someone came to "help" me in this way, would I appreciate it? Would it stimulate me or put me off? Plenary discussion should focus on the question of how to promote the self-management and sustainability of the farmer-experimenter groups. The three basic principles presented on pp. 171–2 may be recalled to summarise this discussion.

Variations:
- To replace the third step, one person or more with long experience of working with the social groups concerned can be invited to react on the recommendations developed by the participants and to share his or her experiences with them.

LEARNING ACTIVITY 2
DEBATE ON THE ROLE OF FARMER GROUPS: A GAME OF OPPOSITES

3–4 hrs

Objectives:
- To recapitulate and synthesise what has been discussed in earlier learning sessions about developing the local capacity for technology development.

- To stimulate participants' thinking about the potential role of farmer groups in technology development and the "price" for individual farmers to realise that potential.

Setting/approach:
- Debate and role play in a workshop setting, with possibilities for small groups to prepare separately.

Materials:
- For visualising concisely the main points from the discussion, preferably cards, markers and a pinboard or large sheet of paper.

Procedure:
- Participants are split into three groups:
  - the "promoters" of a group approach in PTD, who will present the group approach and formulate arguments to support it;
  - the "challengers", who will pinpoint problems with the group approach and critically analyse the promoters' arguments;
  - the "journalists" (a group of 2–3 people), who will formulate what the "public" would like to know about the subject, and who prepare a final overview of the arguments raised for and against the group approach.

- Each group prepares its first statement according to its assigned role (30 minutes).

- One of the moderators acts as chairperson during the debate. The statements (maximum 10 minutes each) are made as follows:

1. the "promoters" explain the group approach to PTD and present their arguments in its favour;

2. the "challengers" draw attention to potential any problems;

3. the "journalists" indicate the issues about which they would like to be kept informed, limiting them to points additional to those already raised by the previous two groups.

- The first two subgroups take about 30 minutes to prepare their second statements, in reaction to the arguments and comments presented and the questions raised in the first round. Meanwhile, the "journalists" try to identify the main issues raised in the first round and the arguments that were given for and against, by writing each issue and argument on a card and then arranging the cards on a pinboard or large sheet of paper.

- The first two subgroups present their second round of arguments (10 minutes each). While these two groups may continue with the debate, the "journalists" withdraw to add the new issues and arguments to the overview they developed after the first round (15 minutes).

- The "journalists" present their summary of the debate and indicate what they personally learned from it. (*Note:* They do not judge the validity of the arguments or indicate a "winner".) The journalists' report may be reproduced and distributed to all participants.

Variation:
- In a simpler form, a discussion may be simulated between a few "promoters", "fieldworkers" and one or two "farmers" who are reluctant to join group activities. Each group prepares its arguments well before the role play starts. Other participants watch the role play and identify the main issues raised, for compilation in the final plenary session.

3 / Farmers' Experimentation • C / Farmer-Experimenter Groups

UNIT 3·D / RECORDING AND ASSESSING EXPERIMENTS

OVERVIEW OF THIS UNIT

EXPECTED RESULTS

This unit will develop fieldworkers' insights and skills for helping farmers document, analyse and assess their experiments, at both community and inter-community level. After completing the learning activities in this unit, the participants are expected:

- to know the basic elements for consideration in planning, recording and assessing experiments;
- to be aware of the complementary role of farmers and their support in recording and assessing experiments;
- to be able to moderate farmer discussions for setting up a recording and assessment system for agricultural experiments.

MAIN CONCEPTS

- **Participatory monitoring and evaluation:** an approach giving the main role to farmers in monitoring and evaluating their own activities.
- **Criteria and indicators** as basic elements in developing a useful recording system.
- **Farmers' criteria** which determine whether they are able and want to apply the innovations.

TRAINING METHODOLOGY

The central issues of this unit can be studied in a "classroom" situation. The design of recording and assessment activities can be studied on the basis of case studies and practised in simple simulations. The difficulties in grasping the essentials of planning such activities should not, however, be underestimated. Very simple examples could be used first (see Learning Activity 1), before dealing with more complex agriculture-based cases. The introductory section on recording and assessing experiments can also be read and discussed by the participants in preparation for the various other learning activities.

LEARNING ACTIVITIES

1. Malika Fernando is ill: Basics of planning a recording system
2. Case study: "No milk for the dairy plant?"
3. Brainstorm: "Outsiders" and "insiders" in evaluating technologies
4. Simulating the setting up of a recording system

DISCUSSION

IMPORTANCE OF RECORDING AND ASSESSMENT

If something is to be learned from experiments, the results must be well analysed. To be able to do this during and after completing an experiment, data must be recorded in some form. In their own experimentation, farmers often (but certainly not always) do this consciously and, after finishing an experiment, can recall information about it in a surprising amount of detail. Studying the existing practices of farmers' experimentation (Unit 3.A, pp. 145–55) will reveal their methods of recording and assessment. This unit gives options to improve on this and to systematise data collection and recording. In fact, assisting farmers in improving their monitoring practices may, in certain situations, be the single most important contribution to strengthening their learning capacities.

Systematic recording is needed especially if:

- the innovation implies a number of important differences, compared with previous practices, e.g. in the use of inputs, such as labour;
- the experiment includes more than one treatment;
- the results are to be shared with a wider audience – farmers from other areas, other agencies, etc.

As PTD focuses not only on developing technology but also on strengthening local capacities to innovate, the monitoring and evaluation activities should cover both of these aspects from the start.

The emphasis in PTD is to support farmers in their own efforts to record and assess the results of their experiments. This section therefore builds on experiences with the Participatory Monitoring and Evaluation (PME) approach (Davis, 1989 & 1990; Rugh, 1986; Stephens, 1988). Questions such as those in Box 3.8 need to be answered, in the first place, by the farmers themselves. The challenge for outsiders is to raise the issues mentioned in the box in words and concepts that make sense to the farmers.

This does not exclude the possibility that outsiders collect and record additional data on their own initiative. This may have the dual purpose of (1) helping to clarify results of farmers' experiments in discussions with farmers and (2) meeting requirements set by the outsiders' professional organisation.

PLANNING A RECORDING AND ASSESSMENT SYSTEM

Planning the recording and assessment of farmers' experiments should lead to a practical list of what to record, how and when. Such planning can be facilitated by trying to answer, one by one, the questions listed in Box 3.8. Data recorded in this way will be easy to process later: going through the list in the box in the reverse order should yield the answers the experiment was expected to give.

FARMERS' CRITERIA

In assessing their experiments, farmers will use various criteria, which may be economic, technical, socio-cultural and/or aesthetic, but generally fall into one of the following two groups:

- Will I be *able* to apply the technology?
  - What claims does this technology put on my/our scarce resources, e.g. cash, labour, water, access to land of certain quality, in what amounts and at what times?
  - What external inputs do I need and will I be able to obtain these regularly and at acceptable costs?
  - What preconditions does this technology require, for example, changes in farm management, co-operation of other family members or/and other families, regular visits of animal health officer, stable output prices, and is it realistic to expect such conditions to be met?

- Do I really *want* to apply this technology?
  - Is it likely to work under my farming conditions? Will I be able to adapt it so that it better suits my personal needs and my particular circumstances?
  - What are the benefits of this technology compared with what I am doing now? How does it help me to realise my priorities or overcome my biggest problems? Does it create any new and interesting opportunities?
  - How certain is it that these benefits will accrue to me, and when? Will these benefits be permanent or temporary?
  - How will others in my family and the community react when I apply this technology? How would this affect me?
  - What other effects will application of this technology have in my farm, within my family, in the wider community, on the environment?
  - Do I like the technology?

Farmers' criteria will vary greatly between households, depending on the productive resources controlled by the household, its social status in the community, etc. But the criteria also vary within a household. The

---

BOX 3-8 KEY QUESTIONS IN DEVELOPING A RECORDING AND ASSESSMENT SYSTEM

What is the OBJECTIVE of the experiment?
What do we want to learn from it?

e.g. to see whether a new cowpea variety is better than the present one, or whether a strong farmer-experimenter group can be established

What CRITERIA should be used to assess an experiment with this objective? This helps to specify the objective. What makes a variety better or a group strong?

e.g. productivity, labour requirements, taste, colour, storability, profits; regular meetings held without outsiders' involvement

What INDICATORS will show whether these criteria have been met?

e.g. in yield kg/ha, or number of meetings per year without project involvement

What do we MEASURE to find the indicators?
What essential information do we have to collect?

e.g. surface of the plot, total production in kg from a plot; or total number of group meetings per year, attendance of project staff at each meeting

HOW do we MEASURE these? How can we collect this information? What techniques of observation and measurement can we use? What equipment do we need? Who will do this, when, and where?

e.g. at sowing, farmers will measure the total amount of seed sown, using local measures (e.g. number of tins); fieldworkers may convert this to kg and relate this to surface area planted

HOW to RECORD data? How do we keep track of what was measured, so that we can refer to it later when we want to compare and analyse results?

e.g. recording forms, wallhangings or calendars, notebooks, farmers' memory

division of responsibilities and tasks is socially defined according to gender and age. This means that different household members will evaluate a technology according to different criteria, which are related to their role and functions in the household.

Criteria to assess a specific technology must be made explicit when screening the technical options prior to experimentation, and can be used again when defining what to record and how to assess the results of the experiments.

COLLECTION AND RECORDING

The information-gathering should be cost-effective for both the farmer and the supporting agency: many things might be interesting to know – especially for the outsider – but not essential for farmers to judge whether the technology meets their criteria. The farmers should be able to understand and use the information gathered. The degree of accuracy needed and its compliance with formal statistical requirements will depend on the type of experimentation farmers and outsiders have agreed upon. What is important is to prevent or reduce biases in observations and errors in measurements as much as realistically possible. It is not always necessary or appropriate to make exact measurements. Orders of magnitude and directions of change are often sufficient to evaluate a technology.

The methods chosen for collecting the information build on methods, measures and expressions used locally. The information should be collected systematically and consistently: participants should have a clear idea about who will do what, when and why. Collection methods need to be flexible: if the information gained is not relevant or useful, the methods should be adapted. Methods that can be applied by farmers to collect information during experimentation include:

- **Farmer record sheets:** experimenters can be assisted in designing simple sheets, or notebooks, for recording the required information periodically (daily, weekly, or each time they do a certain activity). One record sheet is used for each type of information to be collected (e.g. labour input in different operations). Various PTD programmes have given calendars to farmers to note important events. Symbols can be used whenever necessary. In some cases, schoolchildren interview at fixed intervals the other family members involved in the experiments and do the recording.

- **Farmer maps:** farmers can use a map, for example, to record the spread of a certain pest in an experimental and a control plot over time. A farmer can also indicate land-use and management practices on simple maps (e.g. from which hedges the fodder was harvested, when and how much). In a village forestry project in India, the survival rate of different tree species and the influence of sites selected was monitored by placing coloured dots on a map in the appropriate space for each species planted. If the tree died, a circle was drawn around the dot (Stephens, 1988).

- **Board games and other physical tools:** in most cases, creativity is needed to design suitable recording tools. For example, a board game was designed, using pins and symbols for various activities, to enable farmers to record the time spent on each activity (Leesberg & Valencia, 1992). Each day, the household members distributed the rings (each representing a certain amount of time) over the pins. Instead of rings and pins, beans and pots marked with symbols could be used.

- **Group observation and ranking exercises:** periodically, the group members come together on the farm(s) of one, or more, of the experimenters to observe his or her trial and to compare it with their own. Such meetings are preferably held at "critical" stages in the experiment (for example, in variety trials: after crop emergence, or during processing of the product; in a feeding trial, at the end of the season of feed shortage). In comparing methods, discussions can be systematised with the aid of the ranking techniques discussed on p.120. In addition, "open" observations are made and any problems encountered are discussed.

- **Wallsheets to record group development:** recording results achieved may be less difficult than it appears. Groups often use large sheets hung on the wall to record, for example, participation of members in group meetings and joint field activities, contributions in kind or money, training courses taken by members (see also Stephens, 1988).

Notebooks or simple record sheets are used to support monitoring activities

COMPILATION AND ANALYSIS

Information is already collected and interpreted during the experimentation, especially during the above-mentioned joint observations and group meetings. Crucial issues, including farmers' criteria and preferences, are often raised during such meetings. Good reporting is therefore of great importance. However, the collection of some types of information, such as labour inputs, is separated in time from its analysis, which is done after completion of the trial. There will also be a need to bring sets of information and intermediate results together in order to analyse relations between items of information (for example, between costs and effects of the technology).

The PTD facilitator bears in mind that the main purpose of the analysis is for the farmers to learn from the experience. The main challenge is to keep the analysis meaningful for the farmers, rather than primarily for the outsiders. Therefore, methods of analysis have to be developed which the farmers can apply themselves.

The results of an experiment are systematically described and discussed according to the criteria defined during earlier group meetings. To start with, the original objectives of the experiment and the criteria for success are reviewed. The discussion is structured by dealing with these criteria one by one, while maintaining room for additional observations. For each of the main criteria formulated (e.g. yields, weight increase of animals, survival rate), the results obtained *by each experimenter* are noted, and distinction is made between the treatments. The results of all the experimenters can be summarised by calculating averages.

Simple tables can offer an effective way to present this information, for example:

TABLE 3·2
SUMMARY OF RESULTS OF EXPERIMENTS ON SURVIVAL RATES

Experimenter	SURVIVAL RATE OF SPECIES:		
	A	B	C
1	x	xxx	xx
2	xx	xx	x
3	x	xxx	xx
4	xx	xx	xx
average	x	xx	xx

The tables may contain quantitative data, scores (e.g. ranging from three crosses for the best performance to one cross for the worst performance), pictures or symbols indicating certain qualifications (eg. "+" or a laughing face for good quality, "±" for mediocre, and "–" or a crying face for low quality).

The results of the experiment can also be summarised per treatment in similar tables that cover all criteria:

TABLE 3·3
TREATMENT COMPARISON TABLE

CRITERIA	VARIETY A	VARIETY B
No. of roots		
Earliness		
Root rot		
Easy to uproot		
Starch content		

Source: adapted from Ashby (1990)

Subsequent discussion may raise questions such as: Did the soil type make a large difference to the survival of species A, B or C? In some cases, the answer may be obvious. In other cases, a comparison can be made only with the aid of statistical analysis – if the data allow this. In such instances, it is preferable for the need for comparison to be foreseen when designing the experiment. Specialised researchers may provide help in carrying out statistical analysis. Experienced farmer-reseachers, however, may also be interested to learn the procedures, as happened in a PTD programme in Bolivia (Ruddell & Beingolea, 1995). If assistance for statistical analysis is not available, particular care should be taken when designing the experiments, to prevent complications such as those above.

It is important for farmers to identify any special circumstances which may have "distorted" the trials and need to be taken into account when assessing the results. This is seldom forgotten when results are not in line with what was expected, but is equally important when results confirm earlier expectations, because exceptional circumstances may still have influenced the result. Two areas of concern are important:

- Can differences in results between farms be ascribed to differences in the ways in which the farmers implemented the trial? Did something occur that may have strongly influenced the outcome of the trial on one or some of the farms? It may be wise to discuss this in depth, especially when results vary greatly from farm to farm.

- What other general "special conditions" should be taken into account when assessing the results? For example, an extremely dry year or an unusually high market price due to drought. It may be decided to repeat the experiment next year in, hopefully, more representative circumstances.

On the basis of results thus compiled and analysed, farmers will conclude either to reject the technology tried out, or to accept it for wider-scale application, or to experiment further with it, possibly in an adapted form. In view of possibly distorting circumstances such as those mentioned above, one year of experimentation is rarely enough to justify complete adoption of the new technology.

COMPARING "COSTS" AND "BENEFITS"

In many cases, some form of comparison between "costs" and "benefits" will be necessary. Participating farmers can do this, with the aid of the PTD facilitator, by determining, for all operations (from land preparation to storage or marketing) performed in each of the experimental treatments:

- what is needed to carry them out (e.g. for land preparation, you need to hire two person-days of labour and a span of oxen);
- the costs involved (e.g. wages, transport costs, inputs, equipment);
- all effects, including benefits in terms of products, and other positive and negative effects;
- the estimated value of each of these benefits;
- comparison of costs and benefits of each treatment.

Such an exercise can be carried out at various levels of precision. In most cases, it will not be easy to collect information on all the costs involved or to identify and value all the effects. But the exercise is useful even if all costs and effects are not translated into monetary terms: it still provides a systematic framework for review and discussion of the main costs and effects.

Fortunately, it is often not necessary to calculate all expenses and benefits of the treatments being compared; a *partial cost-benefit comparison* can be made. This concentrates only on those aspects that differ between treatments. For example, in a feeding trial, only the cost of the different sources of feed, including labour, may differ and therefore need to be calculated. This greatly reduces the amount of work involved in making the comparison.

EFFECTS ON THE FARMING SYSTEM

During the planning of experiments, assumptions are made about the contribution the innovation may make not only to solving a specific problem, but also to other farming problems and to improving the entire farm system. Therefore, when analysing the effects of an experiment, farmers also need to take account of the development of the whole farm and the wider agro-ecological system, for example, on the basis of the following questions:

- Does this technology help to solve other problems? How?

- Does it create new opportunities for further development of the farm? How? Which opportunities?
- Does the technology:
  - intensify the use of on-farm resources and reduce dependency on external inputs (recycling, enhanced relations between the different productive activities, e.g. crop and livestock interaction)?
  - make the farm more diversified and less susceptible to risk on account of market fluctuations?
  - make the farm less susceptible to drought and/or erosion?
  - contribute to building up soil fertility and increasing the soil's capacity to hold water?

Other questions relevant to analysing effects on the entire agro-ecological system can be derived from the main principles of ecological farming outlined in Unit 1.B, pp. 30–40. Discussion about these issues will deepen the evaluation and contribute to everyone's understanding of the type and direction of changes still needed. This will help in formulating follow-up activities, including the next round of experiments.

PREPARING FOR THE NEXT ROUND OF EXPERIMENTS
Assessment would not be complete without an evaluation of the organisational aspects and planning of the next round of experiments. Aspects to be looked into include:

- What were the strong and weak points in the way we did the experiments this year? What improvements can be made at the individual and group levels? Should collaboration with the supporting agency and other outsiders be improved? If so, how?
- Which results may be valuable for others? For what kind of people or situations in particular? What can we do to inform these other people?
- What experiments should we repeat next year? With what adaptations?
- What other options do we want to try out next year? How do these build on what we learned this year?

LEARNING ACTIVITY 1
MALIKA FERNANDO IS ILL: BASICS OF PLANNING A RECORDING SYSTEM

2hrs

Objectives:
- To develop participants' understanding of the main elements in systematic planning of a recording system.
- To enhance their skills in planning, systematically, such a recording system.

Setting/approach:
- Workshop setting in which participants practise the planning of a recording system in a very simple, non-agricultural case.

Materials:
- Handout 3.2, "Malika is ill", and Handout 3.3 with elaboration of the case.

Procedure:
- The objectives of the exercise are explained, possibly by linking to the need to think more systematically about monitoring in order to prevent too much or too little information being collected. Such a need may have become clear during a previous discussion of participants' prior experiences with monitoring and evaluation.
- In a plenary discussion, the basic questions that need to be addressed when planning a monitoring system are introduced; possibly by reading the relevant section of this unit (pp.177–8). Box 3.8 may be copied for overhead projection.
- Handout 3.2 is distributed or shown on an overhead projector and participants are asked to develop a monitoring system according to the guidelines presented in the introductory discussion. To maintain the speed of this exercise, this could be an individual assignment or a brainstorming session during which participants remain seated and discuss the case in pairs.
- In a plenary discussion, participants share their results for each element of the monitoring system, compare these with the elaboration of the case study (Handout 3.3 distributed or copied for overhead projection), and reflect on possible differences in outcomes.

In a final discussion questions to be raised include:
- the general usefulness of systematic planning;
- the role of planning in participatory planning with farmers;
- the need for specifying details when elaborating how to perform measurements;
- the importance of monitoring side-effects.

Source: PMHE (1993)

HANDOUT 3.2

"MALIKA FERNANDO IS ILL"

Malika Fernando is ill. She is running a high fever, eats very little and does not play at all.

The doctor gives her mother a new medicine, called "Biolife", to treat the illness.

This medicine has only recently been introduced into the country. The doctor therefore urges Malika's mother to note carefully how it works.

HANDOUT 3-3

ELABORATION OF EXERCISE "MALIKA FERNANDO IS ILL"

OBJECTIVE of the "experiment"?
- to be able to judge the effectiveness of the new medicine

Monitoring is also important to stop or change treatment when it goes wrong (side-effects!)

What CRITERIA tell us whether… the medicine is effective?
- if the fever disappears, then the medicine is effective
- if Malika eats normally, then …
- if Malika plays normally, then …
- if there are no negative side-effects, then …
- if …

Which INDICATOR tells us about… disappearance of fever?
- body temperature

WHAT to MEASURE to determine… body temperature?
- skin temperature
- internal mouth temperature
- other …?

HOW to MEASURE… internal mouth temperature?
- use thermometer (equipment)
- thermometer back to zero, insert in mouth 1–3 minutes, read temperature (operation/use of equipment)
- do this 3 times a day during first 2–3 days, then once a day until the fever disappears (procedure)

How to RECORD measurements/data on… mouth temperature?
- notebook
- sheet on the wall to enable Malika's mother to follow development of fever

Source: PMHE (1993)

LEARNING ACTIVITY 2
CASE STUDY: "NO MILK FOR THE DAIRY PLANT?"

2 hrs

Objectives:
- To create awareness of differences in the assessment criteria applied by different household members.
- To draw attention to the limited ability of outsiders to determine "acceptability" of certain innovations, if all household members directly or indirectly affected by the technology do not take part in the assessment.

Setting/approach:
- In a workshop setting, small groups analyse the case study from Nigeria.

Materials:
- Handout 3.4 with case study; additional information, separately, for trainers. For the variation, cards and markers to prepare a problem tree.

Procedure:
- Participants read and discuss in small groups the case study "No milk for the dairy plant?" (Handout 3.4).
- In a plenary session, the trainer helps the participants to collect on the blackboard possible reasons for the failure to increase milk production, as expressed during the groupwork.
- Trainers stimulate further analysis of the reasons given by asking for whom these reasons are most relevant; they can use the additional information provided for them (Box 3.9) to guide the subsequent discussion, focusing on:
  - the effect of the division of labour and responsibilities between males and females, which influences the criteria with which they evaluate certain innovations;
  - differences in access to and control of productive resources and revenues, which also influence those criteria.

Variation:
- To enhance participants' understanding of the reasons for failure, they may be asked to arrange them in a problem tree of causes and effects. To do this, each reason is written on a card and the cards are moved around on the board until the desired cause–effect relationships have been established. This would take at least an additional hour.

HANDOUT 3·4

Case Study
"NO MILK FOR THE DAIRY PLANT?"

Fulani cattlekeepers are the major milk suppliers in Nigeria. Increasing numbers of these pastoralists are settling in the subhumid zone and have taken up some cropping, but cattle husbandry remains their main activity. In the area where they have settled, the annual rainfall of about 1300 mm is concentrated in 6 months, resulting in strong seasonality in forage quality and, thus, in milk yield.

Milk offtake for human consumption is 0.7 litres per cow per day on a year-round average. The milk is sold mainly as fermented skimmed milk mixed with cooked cereals. Butter is also sold, as well as small quantities of soft white cheese.

Earnings from milk products make up about one third of total cash income from the cattle herd. Sales of animals and, to a lesser extent, manure are the main sources of cash for the Fulani households.

Most of the cattle are owned by men and boys, but a few belong to women and girls.

The government launched a programme aimed at increasing milk production. This encompassed:

- establishing milk collection centres and milk processing plants;

- promoting supplementary feeding in the dry season with agro-industrial by-products (e.g. cottonseed cake) or small improved legume-grass pastures ("fodderbanks");

- supplying credit for this purpose, with repayments to be deducted from milk payments at collection centres.

The programme did not reach the set targets. The Fulani did not sell their milk to the collection centres. Although some Fulani bought supplements and/or established fodderbanks in order to provide their animals with additional feed, milk offtake from the herds did not increase. The investments in supplementary feeding were made with the earnings from cattle sales, not with credit or milk earnings.

What do you think were the reasons why this dairy scheme did not work?
What could have been done to prevent such failures?

Source: Waters-Bayer (1986)

BOX 3·9 – LEARNING ACTIVITY 2

BOX 3·9 CASE STUDY: "NO MILK FOR THE DAIRY PLANT?" – ADDITIONAL INFORMATION FOR TRAINERS

The main reason for the poor record of the dairy scheme was the planners' lack of knowledge about traditional milk production, processing and marketing:

- Among the settled Fulani, the men manage the herd and milk the cows. Thus, the men control how much milk is extracted and how much is left for the calves.

- The women receive milk from the milkers (husband or son). Each woman decides how much of the milk is kept for her section of the household, how much is sold and in what form. Women are in charge of all milk processing and marketing activities, and they control the milk earnings.

- As the men are responsible for obtaining herd inputs but the women have the rights to the milk income, the men are not in a position to buy inputs on credit if repayments are to be deducted from milk earnings.

- The traditional milk products are well adapted to local tastes and climatic conditions. With their own products, the women earn 3–5 times more than the price offered for fresh milk by the collection centres. Most women see no need to sell to the centres, even if the herd gave more milk, as they feel they can process and sell much more milk than they are now receiving.

- Women are reluctant to pay for herd inputs, including those which could increase milk yield, as they have no guarantee that the men will allocate more milk to them. They prefer to invest their milk savings in small stock, with the aim of accumulating enough capital to buy their own cattle.

- Men's investments in herd inputs are aimed primarily at maintaining the number of cattle in the herd, e.g. through higher calf survival, so that they have more animals to sell when necessary. Although the cows produce more milk as a result of the supplementary feeding, the men prefer to leave more milk for the calves, rather than giving more milk to the women.

- Analysis of the traditional system with both the men and the women would have revealed how milk processing and marketing could have been improved more efficiently, e.g. by developing small-scale labour-saving technologies for processing grain and milk, by helping the women gain access to clean water for milk processing, by improving transport facilities so that the women can sell their products at larger markets.

- The innovations to increase herd productivity should take into account the patterns of resource control and the division of responsibilities and labour between men and women in the cattlekeeping households.

LEARNING ACTIVITY 3
BRAINSTORM: "OUTSIDERS" & "INSIDERS" IN EVALUATING TECHNOLOGIES

1½ hrs

Objectives:
- To develop participants' understanding of the complementary contributions of "insiders" and "outsiders" in evaluating technologies.
- To reflect on outsiders' role in recording and analysing farmers' experiments.

Setting/approach:
- Plenary discussion prepared for by participants in pairs.

Materials:
- A large blackboard or sheet of paper, cards and markers.

Procedure:
- On a pinboard or large sheet of paper, a matrix is drawn with two columns (headings: strengths & weaknesses) and two rows (headings: insiders & outsiders).

- Participants are given about 30 minutes to discuss in pairs what they see as (potential) strengths and weaknesses of insiders and outsiders in recording, analysing and assessing experiments with new agricultural technologies, to write their findings on cards (one point per card) and to put these in the appropriate box in the matrix.

- The strengths and weaknesses of insiders and outsiders in evaluating technologies are discussed with the aid of the cards in each box.

- The final discussion focuses on the consequences of this for the outsiders' role in recording and analysing trial results, and how these activities should be organised.

LEARNING ACTIVITY 4
SIMULATING THE SETTING UP OF A RECORDING SYSTEM

4–6 hrs

Objectives:
- To develop participants' skills in facilitating farmers' decision-making about the kind of recording and assessment activities they want to do.
- To enhance participants' understanding of important considerations in designing relevant and realistic recording and assessment activities.

Setting/approach:
- Simulation game in which, through various steps, "fieldworkers" help "farmer-experimenters" to design a suitable recording system. The participants should already be familiar with basic issues in the systematic planning of recording systems and with moderating group discussions.

Materials:
- Handout with simple experiment(s), preferably concerning topics that interest farmers in the working area, and assignments for the various actors. An example is provided in Handout 3.5. Large newsprint sheets and markers for reporting groupwork results in the plenary session.

Procedure:
- After a short plenary discussion on important considerations regarding the type and quality of information to be gathered (see discussion section of this unit, pp.177–8), participants form working groups of 5–8 people and are given the handout describing an experiment and the assignment for groupwork.

- In each working group, 4–6 participants act as farmer-experimenters while 1 or 2 others are asked to facilitate the farmers' discussion on recording and analysing the proposed experiment. Before the discussion starts, both facilitators and farmers prepare themselves. It is important that the trainer help the "farmers" prepare well, to ensure that they play the roles realistically.

- In the discussion of farmer-experimenters, the following will be defined for the experiment given:
  - *criteria:* what criteria will the farmers apply in this case?
  - *indicators:* what indicator tells them whether each of these criteria is being met?
  - *measuring methods:* how, by what means, when and by whom could the information needed be gathered?
  - *recording of data:* how, and by whom?

  Trainers may like to prepare possible answers that can be suggested, to give some support to the small groups if they get stuck.

- After the discussion is over (or, at the most, after one hour), the "farmers" give feedback in their working group to the "facilitators" on the latter's behaviour during the simulation.

- The working groups present their findings on sheets of paper (e.g. flipcharts), which are attached to the wall. The groups successively read and discuss, among themselves while walking around, the recording and assessment activities proposed by each of the other groups. They write their comments and questions on sheets attached below or beside the relevant posters with the proposals (comments should take the form mainly of "suggestions for improvement" and "alternative ideas" on what/how to measure).

- In a final plenary session, the comments are briefly reviewed in relation to what was mentioned in the introduction to this learning activity.

Variations:
- Various ways to simplify this learning activity may be considered if the subject is relatively new to the participants. Instead of working out a recording system for all the criteria listed by the "farmers", the group discussion could select to work on only one of them. The feedback about the role of the "facilitators" could also be included in the final plenary session.

- Instead of focusing on an agricultural experiment, a monitoring system could be worked out for an institutional aspect, for example, group development; in such a case, the "farmer-experimenters" should prepare themselves very well, to be able to speak as farmers and not as fieldworkers.

- To hold the discussion with the "farmers" in a truly participatory way, it is best to visualise the main points in some way; trainers may suggest several possibilities to the "facilitators" in advance; if visualisation is relatively new to most participants, it could be introduced first through an additional learning activity.

Source: PMHE (1992)

HANDOUT 3·5 – LEARNING ACTIVITY 4

Case Study
FARMER GROUP "JOINING HANDS"

The farmer group "Joining Hands" is just over one year old. During this year, it has been in contact with the agronomist of a nearby development project.

The farmers' group has asked the agronomist for seeds of a new chilli variety. In line with the project's philosophy, the agronomist explains that he is prepared to provide the seeds on the condition that growing them be used as an opportunity to find out how suitable the new variety is for the local conditions. Some kind of experimentation and data collection will therefore be needed. The farmers' group readily agrees to this, having already had it in mind.

THE EXPERIMENT DESIGNED

The farmers agreed that, each in his or her own farm, they would do a simple experiment, sowing one bed with the new variety and comparing it with a bed of the same size containing the old variety. The issue of data collection was also raised. The farmers felt that they could do this, as long as they knew how to go about it. An appointment was then made for an afternoon meeting of the group, moderated by project staff, to design a simple monitoring system that could be used by the farmers themselves.

THE TASKS

In the afternoon, interested members of the farmers' group met with project staff to design jointly a system of data collection so that they could find out, by the end of the season, which of the two varieties was better under their conditions.

Source: PMHE (1992)

SPREADING AND CONSOLIDATING THE PTD PROCESS

PART 4

PART 4 / SPREADING AND CONSOLIDATING THE PTD PROCESS

4·A / FARMER-TO-FARMER EXTENSION AND TRAINING	■ Farmers spreading results and processes	192
	■ Building on local mechanisms for communication	193
	■ Farmer-extensionists	194
	■ Cross-visits	196
	■ Farmers' training and extension materials	197
	■ Supporting farmer-based extension	198
4·B / SUSTAINING THE PROCESS	■ Sustaining the process and phasing out	207
	■ Strengthening individual capacities	208
	■ Village institutional development	208
	■ Local information bases	209
	■ Horizontal linkages	209
	■ Strengthening linkages with support organisations	210
	■ Monitoring the capacity to innovate	210
	■ PTD organisations as resource centres	212
	■ Monitoring the impact on the agro-ecology	213
BOXES	4.1 Advantages and disadvantages of farmer-based extension	193
	4.2 Roles and activities of farmer-extensionists	194
	4.3 Qualities of an effective farmer-trainer	195
	4.4 Incentives and disincentives for farmer-extensionists	195
	4.5 Possible format for analysing the local spread of innovations	200
	4.6 "The peasants' message"	211
TABLE	4.1 Examples of indicators for process monitoring	211
HANDOUTS	4.1 The Barangay Scholar Programme	202
	4.2 Village production of communication media in El Tigre	206
FIGURE	4.1 A Venn diagram to analyse local institutions	209

UNIT 4·A / FARMER-TO-FARMER EXTENSION AND TRAINING

OVERVIEW OF THIS UNIT

EXPECTED RESULTS

As part of most PTD activities, but especially after community-based experimentation has yielded positive results, farmer-to-farmer exchange and extension and training activities will be carried out. These activities are aimed at spreading the PTD process and promising technologies and attempting to develop an inter-village PTD network. Ultimately this will strengthen the farmers' self-management in technology diffusion and training.

After completing the learning activities of this unit, the participants are expected:

- to understand the main features of farmer-based extension in its complementarity with institution-based extension services;

- to be aware of the importance and the various forms of local communication channels;

- to understand the most common farmer-based extension methods and approaches and the various ways in which they can be supported;

- to be able to reflect critically on their own changing roles at this spreading stage of the PTD process.

MAIN CONCEPTS

- **Local channels and mechanisms for communication:** Local ways that ensure that information spreads within and beyond communities without involvement of outside agencies.

- **Farmer-extensionists:** Farmer "leaders" who emerge from the first experimentation activities and take up the task of training other farmers because of their interest and capabilities.

- **Cross- or exchange visits:** Activities in which small groups of farmers visit each other to observe, discuss and practise particular activities.

- **Farmer training and extension materials:** Materials to facilitate farmer training and extension, developed in close collaboration with farmers.

- **Phasing out of outsiders' support:** PTD practitioners gradually change their role in villages from direct facilitator to consultant and advisor.

TRAINING METHODOLOGY

In discussing the issues of this unit, one can start with a critical reflection on participants' experiences with extension approaches of different organisations in their area, and their strengths and weaknesses. This will help participants to understand the complementarity between the approaches and the farmer-to-farmer extension activities being carried out. This analysis will be much stronger when participants are asked to interview farmers on their experiences. Such farmer interviews are especially important in studying local communication channels and their relevance. They also provide an opportunity for participants to practise the preparation and use of semi-structured interviews.

This unit approaches training in farmer-to-farmer extension methods and techniques by providing several case studies that will be used for joint analysis. Where possible local case studies, if available, have preference, if only because people involved may be visited and interviewed. The discussion section of this unit gives a general overview of the relevant issues and may be read and studied at appropriate moments.

LEARNING ACTIVITIES:

1. Farmers' views on extension approaches
2. Understanding community communication channels
3. Case study: Farmer-based extension
4. Developing guidelines for selecting effective farmer-trainers
5. Organising cross-visits
6. Preparing for local media production

AUDIOVISUALS

- Farmer-centered extension in the Philippines. Slide–tape programme by IIRR (see Appendix, p. 222).

DISCUSSION

FARMERS SPREADING RESULTS AND PROCESSES

In conventional agricultural extension, the emphasis is on spreading the use of agricultural technologies. In PTD, however, the sharing of such technologies, which are the locally realised *outcomes* of farmer experimentation, is only part of the challenge. A major emphasis will be on sharing with other communities the *basic ideas and methods of how to identify, test and adapt promising technologies* to develop more sustainable agricultural practices. Giving farmers an important role in these sharing activities, as seen below, does not in itself guarantee sufficient attention to these process-related aspects. One should be careful with concepts such as "farmer-demonstrator" or "contact-farmers", which often indicate a focus on teaching technologies.

In many of the previously described PTD activities, there are situations where farmers learn from other farmers. Farmers may indeed play an important role and take over responsibility in spreading experiences on agricultural innovations. This is known as farmer-based extension. Box 4.1 lists important advantages and disadvantages of this approach. Institution-based (NGO and GO) extension plays a complementary role, both in terms of supporting farmer-based extension and facilitating extension beyond the capacity of farmer-based systems.

BUILDING ON LOCAL MECHANISMS FOR COMMUNICATION

Spontaneous diffusion of technologies (that have been proved successful) occurs frequently when ideas are shared with friends, seed materials are exchanged, and new products gain recognition along trading routes. Local markets or meetings may be important venues for sharing agricultural ideas. A great variety of methods – drama, songs, jokes – may be important locally to carry agricultural messages. There is now increasing recognition of the importance of indigenous communication networks (Box, 1989; Simpson, 1994); also within the context of PTD programmes (Gubbels, 1988; Budelman, 1983).

Such informal communication networks may, however, have their limitations: information is often shared haphazardly if or when the opportunity arises, limited resources may prevent exchange beyond the local neighbourhood, and gender or other socio-cultural conditions may be equally restricting. A survey among women farmers in Burkina Faso, for example, revealed that only one per cent of those with knowledge of certain new crop technologies had learnt this from their husband (Saito & Spurling, 1992). The value of information from local channels may also be discredited over the years through the heavy emphasis on new ideas spread by formal extension.

BOX 4-1 ADVANTAGES AND DISADVANTAGES OF FARMER-BASED EXTENSION

ADVANTAGES:
- The farmer-extensionist is familiar with the local characteristics, problems and history
- S/he speaks the language of the farmers and understands them
- S/he knows how to motivate neighbours
- S/he has good contacts and friends within the community
- People have more trust in someone from the same group who is actually delivering or initiating the technology him/herself
- Farmer-extensionists are used to manual work and walking long distances
- The costs of maintaining farmer-extensionists are much lower compared with those of outsiders
- Farmer-extensionists can show others what they have accomplished on their farm

DISADVANTAGES:
- The farmer-extensionists may neglect extension work, as they live where they work and have many other responsibilities
- Some are reluctant to learn from a local person; better-off people may look down on farmer-extensionists as they may be less well off
- Farmer-extensionists may require more training, and sometimes learn slowly
- Frequent trips outside of the area may cause family problems
- Farmer-extensionists may have difficulties in preparing reports and other paperwork

Source: Mag-uugmad (1994)

In the spreading of experiences and results from farmer-based experimentation, PTD programmes attempt to build on the existing information networks and to address some of their limitations. Important elements in such efforts are:

- **Identification** of informal communication and diffusion channels.

> **BOX 4-2 ROLES AND ACTIVITIES OF FARMER-EXTENSIONISTS**
>
> **TOWARDS THE COMMUNITY:**
> - Facilitate problem identification
> - Provide technical assistance and training
> - Look for resources both within and outside the community
> - Facilitate experimentation and evaluation
> - Look for the necessary information
> - Help to plan and organise activities
> - Support local leadership in development initiatives
> - Facilitate monitoring, evaluation and follow-up of community projects
>
> **TOWARDS THE SUPPORT ORGANISATION:**
> - Participate in planning activities with support organisation
> - Act as facilitators and guides in training activities and field-trips provided by the organisation
> - Provide information for progress reports
> - Estimate the resources needed for project activities
> - Provide links for existing knowledge and resources
> - Make staff of support organisation aware of the community's real needs
> - Coordinate activities with other organisations working within the same community
> - Facilitate communication in local language, act as translator
> - Act as channel of communication (and interpretation!) between community and support organisation(s)
>
> *Source: adapted from Selener et al. (forthcoming)*

- **Direct "use"** of these informal channels; utilising traditional village-level social processes (exchange of information at the market place, when fetching water, etc) and modes of communication (folksongs, storytelling, folkdrama).

- Strategically **identifying village clusters** on the basis of existing communication linkages and working with a limited number of "motor" villages: here the PTD process is initiated to spread later to the whole cluster of villages.

- **Direct support** to farmer-to-farmer communication and training. Building of a farmer-extensionists' network, facilitation of cross-visits, and support to the development of farmers' manuals and audio-visuals are among the best-known approaches in this area and are described below.

FARMER-EXTENSIONISTS*

Building a cadre of local extensionists is a logical step in many PTD programmes when potential farmer-trainers emerge from among the first generation of farmer-experimenters. They can be complementary to existing institution-based services but can be especially important in the growing number of areas that are not reached by these formal services. Crucial issues for consideration in setting up such activities include: formulation of the tasks and roles of farmer-extensionists; selection process and criteria; institutional setting and remuneration; appropriate training and support services.

The *role* of a farmer-extensionist is essentially to teach other farmers in his or her community (or in other villages) the technologies s/he is successfully adapting/practising, and to encourage others to experiment in similar ways with these and other ideas. In practice, his or her role and activities may include much of what is mentioned in Box 4.2.

To make the *selection process* of farmer-extensionists a joint process, it is good to develop with (groups of) farmers a list of criteria of what makes a good farmer-extensionist, such as those in Box 4.3. Of course, such criteria can vary depending on socio-cultural context and the list should therefore be treated with care. Although some programmes start their work almost immediately with the selection of farmer-promoters or farmer-trainers, it is often more effective to do this after one activity or more has been implemented. These activities often lead to the emergence of people with an interest in, and the capacity for, training others. Also at this stage, the farmers and staff from support organisations have started to know each other better and this should enable good communication on the expectations of the farmer-extensionists.

In line with the general PTD approach, communities or farmer groups are often the ones to select the farmer-trainers. The role of the PTD practitioners is to raise certain issues and to add criteria; for example,

---

* This section benefits greatly from Selener *et al.* (forthcoming)

> **BOX 4-3 QUALITIES OF AN EFFECTIVE FARMER-TRAINER**
>
> - S/he has adapted and successfully established the technology.
> - S/he has gone through a process of experimentation on his or her farm.
> - S/he is older, respected, influential.
> - S/he is not too caught up with other (e.g. family) responsibilities, or is able to solve possible constraints – in other words, has time.
> - S/he is stable (not always travelling) – is available.
> - S/he is credible, has good relations with the community.
> - S/he shows interest, and has volunteer spirit to take up the role.
> - S/he is honest/humble and does not only think of his or her own welfare, has a sense of commitment.
> - S/he is willing and capable to train other farmers.

to recommend the selection of women as farmer-trainers. It may also be the role of the PTD practitioners to propose selection procedures to counterbalance the possible over-influence of the local élite. Voting by (small) groups, rather than by individual farmers, has been suggested in this context (Joel Zwier, 1995, pers. comm.)

Care should be taken in the selection process to prevent people being chosen not on the basis of their qualities but because of political considerations (e.g. a person may hope to get a paid job). Often, female farmers are not recognised as effective peasant farmer-trainers (which may be remedied through dialogue) or socio-cultural norms restrict their mobility. This still does not exclude women from functioning effectively as farmer-trainers, but more creativity may be needed to organise their selection.

In defining the appropriate *institutional setting* of the farmer-extensionist, questions that need careful consideration include:

- Will farmer-trainers receive remuneration for the work they do, or is it solely volunteer work? Often the first option is chosen in an effort to develop more structural solutions. As a rule of thumb, the remuneration should be equal to the wage of a day labourer, plus travelling expenses. In some programmes, the remuneration is twice that of the going rate for hiring a day labourer. But who will provide funds for this in the long run?

- To whom are they accountable? To local community structures, a farmer organisation, the support organisation? The first two options, if available, seem to have the advantage of direct accountability to the people involved. Nevertheless, farmer-extensionists frequently end up becoming part-time NGO staff members. To what extent are these organisations accountable to their target groups?

> **BOX 4-4 INCENTIVES AND DISINCENTIVES FOR FARMER-EXTENSIONISTS**
>
> **MOTIVATIONAL INCENTIVES:**
>
> - Appropriate salary
> - Allowance for transportation and food
> - Training and field visits to other projects
> - Participation in planning, evaluation and decision-making
> - Recognition of good-quality job from NGO and community
> - Fringe benefits and perks (medical insurance, bonus)
> - Visits to his or her project area by outsiders
> - Provision of technical books, pamphlets, etc and other resources
> - Certificates when attending courses, seminars, etc
> - The replication of his or her work in other communities
> - Respect from extension agents, NGO staff, etc
> - Work with motivated communities
>
> **DISINCENTIVES:**
>
> - Low salaries; they are paid much less than other NGO staff who do the same job
> - Lack of training
> - Lack of promotion or equal opportunity
> - Lack of decision-making power within NGO
> - Lack of responsibility or motivation by the community
> - Unjustified complaints from NGO or community
> - Lack of trust and unrespectful treatment from fieldstaff from NGO
> - Racial and class discrimination
>
> *Source: Selener et al. (forthcoming)*

- Are farmer-extensionists expected to work within their own village or to serve outside it? Where the former is the case, they are often volunteers who have been chosen by their community. In the latter case, they are recruited by the support organisation either directly or together with the community.

- How many farmer-extensionists are to work in each village? In general it seems best to have as many, with the right experience and qualities, as can be trained and supported. This often leads to a certain specialisation among them, towards e.g. soil conservation next to animal husbandry and/or pest management. The question then arises: who will take care of process-related activities, supporting problem analysis, or organising meetings. Is there a need for a generalist? All these questions are familiar in the planning of institution-based field extension.

Finally, adequate *training and support services* need to be put in place. Very revealing in this context is the list prepared by farmer-extensionists in Latin America of what motivates or demotivates them (Box 4.4).

CROSS-VISITS

One of the most effective farmer-based methods to extend the process and technology to new farmers and new villages is cross-visits or farmer-to-farmer visits. Different *types of cross-visit* can be distinguished. One type is to take farmers from a new project site to visit farmers in an established programme area as a means of quickly launching the process and technologies in the new area. Another type is village-to-village exchange within the same project area or exchange between farmers in one village. This is to encourage sharing of experiences and mutual learning. Usually visits involve a limited number of farmers (2–6), to enable intensive interactions. In the case of larger groups, enough opportunity should be created for interaction within smaller groups.

If the *aim* of the cross-visit is to provide motivating information, it can be relatively short (1 day); specific training is provided later. If the aim is motivation and training, the visit should last 3–5 days to permit "learning by doing" in real conditions: actually practising the new technologies by working with the host farmer in his or her fields.

Farmer cross-visits play a number of very important roles, several of which often go unnoticed:

- Cross-visits permit farmers to learn of new technologies and possible adaptations of these to their own situation.
- They enable farmers to see the result of technologies which they have not yet used.
- The farmer who is visited gains moral support from sharing his or her experiences.
- Longer-term linkages between farmers and communities are established which serve as a basis for future agricultural development.
- Cross-visits often result in a transfer of new materials: seeds or examples of new tools and equipment.
- Cross-visits expose delegates to the concept of farmer-extensionists and the process of farmer-led experimentation.
- Visits to other areas and exposure to new situations may help to strengthen the self-confidence of delegates, thus helping to develop local leadership.

The *role of the fieldworker* in organising a cross-visit includes the following:

- to introduce and discuss the idea of a cross-visit;
- to provide help in choosing delegates (the guidelines for selecting effective farmer-extensionists already mentioned also apply for guiding the communities to choose good delegates, including the maintenance of proper balance in terms of age and gender);

Cross-visits are one of the most effective farmer-based extension methods

TRAINING MATERIALS DEVELOPED BY FARMERS FOR FARMERS

BASIC METHODOLOGY KIT:

1 spirit level, 1 ball of string, litmus paper,
1 tape measure, 1 calculator,
1 set permanent colour markers, 1 roll adhesive tape,
1 VHS video-cassette (with 6 sociodramas
and a video clip), 1 pencil, 1 ballpoint pen,
1 notebook, 1 weighing scale, 1 wooden chute,
1 wooden A-frame, 1 training manual, the books
Tierra Fresca ("New Ground"), *Sembrando Futuro*
("Sowing for the Future") and an annual field agenda

SOURCE: SIMAS (1995)

- to assist in clarifying objectives: link the visit with an identified problem or with intended experimentation with a certain technology; is the focus mainly on motivation/information, or on learning/training?

- to facilitate, where necessary, the cross-visit's enabling of direct farmer-to-farmer interactions (which entails reducing the profile of the inevitable officials); this can be facilitated by specific arrangements such as electing a spokesperson for the group, giving money for the transport to another group member, and distributing questions on certain topic areas among delegates in advance;

- to encourage and help organise "de-briefing" after the cross-visit to assist delegates in summarising the major points of interest and deciding on a plan of action: e.g. how and what to report back to their respective communities, what can be applied;

- to monitor the results of the visit, following up decisions made after it.

FARMERS' TRAINING AND EXTENSION MATERIALS

Documentation by farmers of their experiences in PTD activities plays a very important role in spreading the ideas; it helps to make these experiences accessible for other villages and areas. The documentation process in itself helps to reflect on and analyse in a systematic way what has been achieved, thus enhancing the self-management capacity of the farmers. Many exciting opportunities exist:

- Local means of communication may be used and/or reinforced: songs about successful activities are sung in many villages, in schools or at social occasions; drama is also often used (e.g. Adoyo, 1994)

- Farmer-extensionists may be assisted in producing audiovisual recordings of experiences: taping farmers telling each other about the history and results of an experiment; "photonovellas" – regular photo-based journals – are another example developed by Campesino a Campesino in Nicaragua;

- Visualised results of activities (maps, monitoring sheets, harvest data) may be kept and stored for use in various situations and interactions;

- Experiences may be shared through quarterly farmer journals or compilations of basic farmer-training materials (see above).

Development and production of many of the above media can be completely in the hands of the farmers or communities themselves, although in certain cases outside technical expertise will need to be "hired in". Sometimes an outside group may take the lead in developing materials for later use by farmers. Strong interaction with farmer groups is always required, to ensure the relevance and user-friendliness of the materials developed.

Whatever the situation, it is important to understand that these communication media are used in a different way and play a different role in farmer-based extension, compared with conventional extension approaches. In the conventional transfer of technology, communication media are meant to assist the extensionist in putting a message across – to transfer information on varieties and practices recommended by scientific research, and to convince farmers of the advantages of the recommendation so that they will follow it up. In farmer-based extension, the media are meant to assist farmers in analysing their experiences and sharing them with other farmers. It will be the farmer who produces the "message" and the aim is not to "convince" but to enable other farmers to make adequate decisions and to try things for themselves. This basic difference strongly influences how media are produced and used in a PTD programme.

SUPPORTING FARMER-BASED EXTENSION

Defining adequate support activities is a major challenge to PTD programmes: how to support without taking away the initiative and sense of ownership from farmers, groups or communities? The following aspects need careful consideration:

- **Facilitation:** this "magic" word, already frequently mentioned, encompasses all efforts to stimulate direct farmer-to-farmer interaction; it implies being present and supportive without being the centre of attention and only for as long as is really necessary;

- **Encouraging the participation of all:** going beyond facilitation, PTD practitioners may put on the agenda participation of the poor, of women or of other specific groups and suggest ways that ensure their involvement;

- **Training support to farmer-extensionists:** regular input in discussing certain technologies, possibilities for and ways of experimentation, moderation and training skills, leadership aspects, community organising, etc. Apart from the content of the training, the training approach itself is an important learning opportunity: farmer-extensionists have indicated that exposure to participatory, highly interactive, adult education methods has strongly influenced the way they work with other farmers (Zwier, 1995, pers. comm.);

- **Support in the development of training and extension materials:** provision of certain technical skills as well as helping farmers to develop these skills themselves;

- **Financial support to honorarium:** especially when farmer-extensionists are expected to serve farmers in other villages, their work cannot always be entirely supported by local resources; the extent of and conditions for honorarium support require careful discussion with the farmers (groups) involved;

- **Provision of critical resources:** this may include money for transport, the purchase of a simple slide projector, or subscription to a relevant local or national journal;

- **Monitoring support:** keeping track of the results and impact of farmer-based extension activities for discussion with the people involved; encouraging evaluative meetings among farmer-extensionists.

The emergence of a cadre of village extensionists and the spread of activities to a greater number of villages has important implications for the role of the PTD practitioner. There is a development from a frontline facilitator and liaison officer in a few initial villages to an orchestrator of a larger process. S/he will gradually withdraw from the role as an intervening actor in the first-generation communities, and the experimenting farmers themselves will have to secure the continuation of the local experimentation process. The initial group of farmers will also have to take on (part of the) responsibility for the diffusion of the developed technologies and the experimental process to other villages. In the initial villages, the PTD practitioners will increasingly become external consultants and supporters of the farmer-extensionists. Consciously managing this shift is a major challenge for fieldworkers in this stage of the process.

An important aspect of the new role of the fieldworker is careful assessment of technologies that merit further promotion, taking into account their potential for the development of LEISA. Subsequently, s/he will identify for whom and where these results may be relevant, taking into account the specific context within which the technologies were developed and related conditions for wider diffusion (socio-cultural, economic and political).

LEARNING ACTIVITY 1
FARMERS' VIEWS ON EXTENSION APPROACHES

3–4 hrs +travel

Objectives:
- To stimulate critical reflection on the extension concepts and methods practised by, or known to, participants.
- To enhance understanding of farmers' (true and false) expectations of agricultural extension.
- To clarify one's own role and views towards farmers' expectations – especially those one does *not* want to live up to.

Setting/approach:
- Participants interview farmers on their experiences with agricultural extension.

Materials:
- To be selected by participants for use during the interview, if any. Newsprint paper and markers, or cards, to present main findings to the plenary.

Procedure:
- Introduce the subject and objectives of this exercise.
- Ask the participants to form small groups (2–5) for the farmer interviews; each small group is to interview groups of farmers who have experienced different extension situations or approaches. Be aware that what trainers and/or fieldworkers define as "different" might be viewed by farmers as "the same", and vice versa.
- Give the groups an opportunity to prepare themselves. Note that this is another chance to practise semi-structured interviewing; an important element is the preparation of a checklist with key questions, covering probably three main elements:
  - *description:* e.g. what does the extension worker do? what kind of activities? how does s/he act in this?
  - *evaluation:* e.g. how successful is the extension worker? what do farmers like and dislike?
  - *analysis:* e.g. why do the extension service and the extension worker function as they do? what suggestions for changes do farmers make?
- After they have been analysed in the small groups, the findings of these interviews are presented in a plenary session; a joint reflection on the results may focus on:
  - How do the different "approaches" compare with each other from the farmers' point of view? Do we agree with this view?
  - What are the most important elements in the farmers' expectations towards extension; which ones do we find difficult to accept? How to handle this?
  - What does this imply for PTD programmes?

Variations:
- Combination with Learning Activity 2.

LEARNING ACTIVITY 2
UNDERSTANDING COMMUNITY COMMUNICATION CHANNELS

3–4 hrs +travel

Objective:
- To enhance participants' understanding of local patterns of communication of technological information, and their importance as a basis for developing a farmer-based extension and training system.

Setting/approach:
- Participants moderate a discussion with small groups of farmers in the field.

Materials:
- To be selected by participants for use during the small group discussion, e.g. to visualise main points raised.

Procedure:
- Introduce objectives and procedure.
- Ask participants to form (small) teams and allow for preparation of fieldwork (such as deciding which participants will do what during the group interview and discussion).
- The teams hold a meeting with a small group of farmers. The farmers are asked to indicate one improvement or change they have made in their farm plan or cultural practices in the last few years. The improvements mentioned are listed for all to see (blackboard, large sheets of paper) together with the name of the farmer who made that change (see, for example, format given in Box 4.5). Be aware that the question does not mean innovations offered to them by the NGO or government services; i.e. it should be their own innovation.
- For each of the improvements listed, the farmer who made it explains when and why s/he made the change and how s/he obtained or developed the idea for it (e.g. own idea, developed within the farm household, from discussions with neighbours, or relatives and friends, from others in the village, from government or NGO fieldworker, on the market, in the bus, from the newspaper, television, etc). The main points are again noted on the blackboard.
- For each of these technologies, farmers are encouraged to discuss amongst themselves:
  a) how long this practice has been used in the village;
  b) where the idea originally came from and how it arrived in their community (if from outside);
  c) how it spread in the community (if it has done so).
The main outcomes are again noted for all to see.
- Based on the outcome of the above discussions, farmers and participants could discuss:
  a) the issue that farmers themselves are, in fact, extension workers and often act as trainers;
  b) how farmers could further help other farmers to learn about new techniques and varieties on a more regular basis.
- After the fieldwork, the participants summarise the main lessons learnt and (either directly in plenary or first in small groups) discuss the implications for developing farmer-based extension and training.

Variation:
- Resource persons may be interviewed to share their views on community communication patterns, either in the classroom or elsewhere – teachers at the local school, village people experienced in certain media, anthropologist who has studied in the area on these topics, etc.

BOX 4·5 POSSIBLE FORMAT FOR ANALYSING THE LOCAL SPREAD OF INNOVATIONS

Technology and farmer	When/why was change made?	How did s/he get the idea?	How/when did idea come to the village?	How did it spread?
1. …				
2. …				
3. …				
etc				

LEARNING ACTIVITY 3
CASE STUDY: FARMER-BASED EXTENSION

2–3 hrs

Objectives:
- To increase participants' understanding of the concept and practice of village-based extension.

- To stimulate participants' thinking on how such an approach could be put into operation in their own situation.

Setting/approach:
- This activity provides for the study of a case in small groups in the classroom situation. When local case studies are available, a field visit may complement this.

Materials:
- Handout 4.1 with the case study "The Barangay Scholar Programme" and/or the slide-tape "Farmer-centered Extension in the Philippines"; materials to present the results of the small-group work.

Procedure:
- Introduce the topic and objectives of the activity; distribute the case study on village-based extension and/or present the slide-tape. However, case studies of local organisations should be identified and used whenever possible.

- Ask participants to analyse the case study in small groups along the following lines:

  - identify the major elements of the approach;
  - review how these elements are implemented;
  - discuss the strengths and limitations of the various elements and of the approach.

- Results of the group work are presented briefly; joint reflection on the results may focus on:

  - deeper analysis of the strengths and weaknesses mentioned;
  - the extent to which various elements are controlled by farmers or outsiders, and the reasons for this.

- Finally, participants may be asked in the plenary – individually, in teams or jointly – to summarise the main lessons by outlining the approach they would wish to follow; outcomes of the discussions may be reproduced and distributed or displayed on the wall.

Variations:
- If local case studies have been identified, a field visit may be organised instead of using slide shows or written information.

Source: slide-tape "Farmer-centered extension in the Philippines", produced by IIRR (see p. 222)

4 / Spreading and Consolidating the PTD Process • A / Farmer-to-Farmer Extension and Training

HANDOUT 4.1 – LEARNING ACTIVITY 3

THE BARANGAY SCHOLAR PROGRAMME

The Philippine Rural Reconstruction Movement (PRRM) developed a farmer-based technology dissemination approach, which has proved to be a very efficient and cost-effective method for diffusing innovations at community level. The main elements of the process of development and dissemination are the following:

- After assessment of the knowledge base at (a) the farmer level (indigenous knowledge) and (b) the institutional level, and design of the operational strategy including guiding parameters (useful for M&E), a process of farmer-level testing, adaptation, refinement and adoption is developed with *farmer volunteers* during an initial phase of 12–14 months.

  The intention is to move towards integrated farming. However, the farmer decides what he is going to do, selecting from a range of options, and the technology *per se* is not presented as an integral package. In the initial phase, the focus is on combining two or three technologies which will realise a short-term increase of income (mainly by reducing production costs), even as the long-term goals are slowly being realised.

- Volunteer farmers are visited regularly by project staff to discuss what needs to be done or revised; where required, hands-on training is given by a peasant promoter. It is important to help farmers understand the linkages between various options, and to promote group formation among volunteer farmers. The interactions among farmers encourage them to experiment and to learn from the process. From the initial farmer volunteers in the new project site, about 5 farmers are selected who will visit an old project site in another province.

- Each group of initial farmer volunteers selects a farmer leader and one or more peasant farmer-trainers (Barangay scholar) at the end of Year One. Each group is responsible for conducting local-level training for a wider group of interested farmers at their own site (by farmer leader and peasant farmer-trainer) and to monitor and assist experimenting farmers in that cluster.

- From among the peasant farmer-trainers, competent farmers are hired as promoters with a minimum wage equivalent to 1 day/week. The rationale for paying them is that they can then pay a worker back home. The promoter acts as a resource person and farmer-trainer in one or two new villages in the same area.

- Cross-visits: farmers in one group visit farmers in another cluster to pick up new ideas and to be trained in certain technologies and skills.

- Wider sharing of the experiences with visitors, documentation (case studies, slide shows, newsletters), workshops and training. A special role is played by the production of idea kits (sources of technological options).

HANDOUT 4•1 – LEARNING ACTIVITY 3

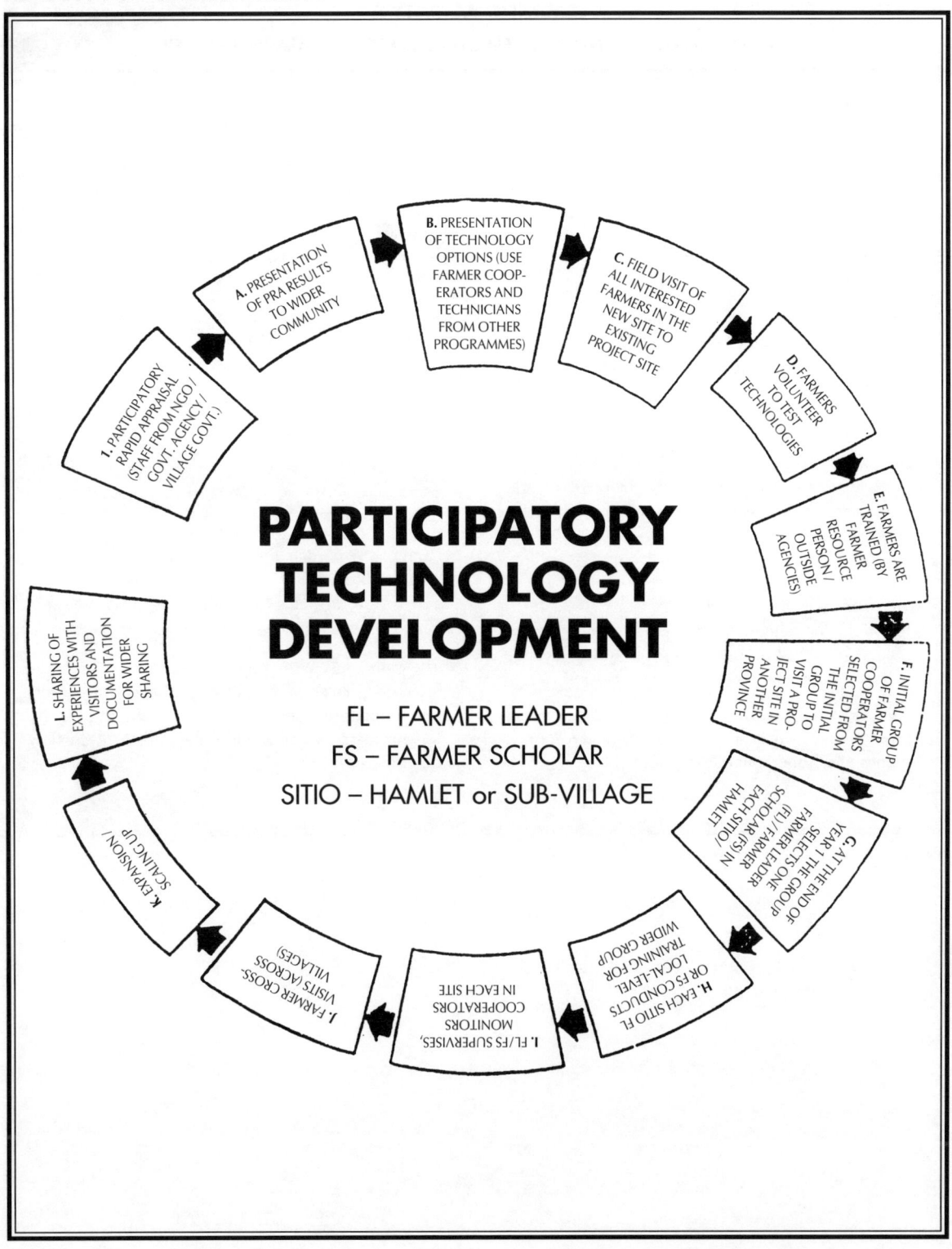

4 / Spreading and Consolidating the PTD Process • A / Farmer-to-Farmer Extension and Training **203**

LEARNING ACTIVITY 4
DEVELOPING GUIDELINES FOR SELECTING EFFECTIVE FARMER-TRAINERS

2–3 hrs

Objectives:
- To understand the role of farmer-extensionists and recognise the qualities required for them to be effective.
- To be aware of major problems or pitfalls in recruiting farmer-trainers and ways to overcome these.

Setting/approach:
- In a plenary workshop setting, participants discuss in pairs key questions on the theme, alternately reviewing and summarising their answers.

Materials:
- Cards and markers, board for cards, pins or tape. Background reading – possibly the relevant part of the discussion section of this unit (pp. 194–6).

Procedure:
- Introduce the training topic in terms of the learning objective, and ask the participants about its relevance to the process of community-based extension.
- Ask the participants to discuss in pairs (buzz groups) for 5 minutes: what should be the role and responsibilities of a farmer-trainer/extensionist?
- Answers are written in keywords on cards. One by one the pairs pin their cards on the board and explain their keywords; repetitive cards are pinned next to each other.
- Moderate a discussion for clarification and synthesis; add new elements or perspectives.
- In a similar way, ask the pairs: given this role, what are the key qualities or characteristics of a farmer-trainer? Reporting and synthesis as above.
- Finally ask the pairs: what problems or pitfalls have to be avoided when selecting farmer-trainers?
- Prepare a list of major problems and pitfalls on the board by asking each pair to report one not yet mentioned by a previous pair.
- Each pair of participants will choose or be given one of the problems listed. Ask them to analyse the problem and to draw on their experience to suggest how to prevent the pitfalls and deal with the problem.
- The pairs present their main suggestions, which are listed on a blackboard or flipchart. These are reviewed and summarised, possibly in the form of a list of issues for consideration in facilitating community discussion on farmer-trainers.

LEARNING ACTIVITY 5
ORGANISING CROSS-VISITS

2 hrs

Objectives:
- To understand the advantages of cross-visits as a method to spread the PTD process and promising technologies.

- To be aware of main points in organising and facilitating farmer cross-visits.

Setting/approach:
- In a workshop setting, guided discussion plus interview with farmers who have participated in cross-visits. In the variation, actual facilitation of a cross-visit.

Procedure:
- Introduce the topic, learning objectives and working procedure.

- In a guided plenary discussion, stimulate participants to draw from their own experiences what they know of cross-visits:

  • what are cross-visits? what types of cross-visit are there? what are the aims of cross-visits?

  • what roles might cross-visits play?

  • what makes cross-visits effective?

  The results of the plenary discussion on each of these points are summarised for all to see.

- On the basis of this, ask participants (possibly in smaller groups) to prepare questions for the farmer interview; invited farmers who have participated in cross-visits are then interviewed about their experiences.

- Participants summarise in their small groups the main conclusions on how to organise and implement a cross-visit; the conclusions are copied and distributed to all participants.

Variation:
- After the training, participants are asked to implement one or more cross-visits in their working area, using the summary of main conclusions from the small-group work as a basis. After some time, participants meet again to exchange experiences, review problems and consider how skills and procedures can be improved.

LEARNING ACTIVITY 6
PREPARING FOR LOCAL MEDIA PRODUCTION

3 hrs

Objectives:
- To broaden participants' view on the use of "media" in agricultural extension.
- To broaden their view on the range and types of media and channels available for the exchange of farmers' experiences.
- To develop guidelines for the use of selected media/channels.

Setting/approach:
- Workshop setting in which small groups work in two rounds.

Materials:
- Handout 4.2 with case study "El Tigre" (or self-made example). Materials to present results of small groups.

Procedure:
- Explain objectives and procedure.
- Distribute Handout 4.2 with the case study "El Tigre" (or self-made example). Ask the participants to study the case in small groups and to make a list of local ways and means of communication and sharing experiences, and other non-conventional ways of communication that may be used in farmer-based extension.
- The lists are displayed and each of the options is briefly explained and its viability discussed. The most promising options are selected by the participants.
- Each group selects one option, and develops a set of practical guidelines, indicating how a fieldworker can use that option to stimulate the sharing of experiences among farmers within and between villages.
- Each group comments on the guidelines produced by one other group; the revised guidelines are distributed to all participants.

HANDOUT 4.2

VILLAGE PRODUCTION OF COMMUNICATION MEDIA IN EL TIGRE

The villagers of El Tigre, Colombia, developed improvements in the construction of the traditional *azotea*, an old canoe put on poles to protect it against the regular floods, which served as a home garden. The members of the group met regularly to discuss progress and problems. At the end of the second growing season, the group reviewed the process they had gone through: from the diagnosis of the problem to the consumption and storage of the vegetables. The discussions were tape-recorded and a local artist made a drawing with the farmers to illustrate each of the steps. This resulted in a simple manual in which El Tigre explains to other villages in the area why and how it conducted its *azotea* experiment. The manual was distributed to the neighbouring villages. El Tigre farmers used the manual to explain the experimental process to visiting farmers and to train these farmers in construction and management of their *azotea* and related husbandry and processing practices.

Other villages used other means to share their experiences with other villages: one group produced a wall paper on their experiment, including *copla* (a traditional form of poem/song) and drawings. The wall paper was copied several times and hung at central places in neighbouring villages. In one instance the results of a group review meeting were illustrated by drawings made by the participants. The project staff transferred the drawings onto a series of slides, which were used by representatives of the farmer organisation to start a similar process in other areas.

Source: based on A. Cardono, L. Orozco & C.I. Mazo (pers. comm.)

UNIT 4·B / SUSTAINING THE PROCESS

OVERVIEW OF THIS UNIT

EXPECTED RESULTS

From the very beginning, attention must be paid to creating conditions that will help make the process of technology development sustainable. Outside facilitators thus facilitate the consolidation and intensification of the PTD process. A major emphasis is on strengthening the role and responsibility of farmers and communities in PTD, building on the mechanisms for farmer-to-farmer extension and training (Unit 4.A, pp. 193–4). PTD support organisations also have a role to play to ensure that the approach gets embedded in local agricultural development programmes and initiatives.

After completing the learning activities of this unit, participants are expected:

- to be familiar with strategies and mechanisms to ensure development of local capacities to continue PTD and to recognise their own role in this process;

- to be able to reflect critically on their role in the development of direct linkages among farmer groups and communities and between farmer groups and support institutions;

- to be able to develop their approach for monitoring the development of local capacities to innovate;

- to understand the importance of documentation in consolidating the PTD process and to be familiar with several practical ways to do so.

MAIN CONCEPTS

- **Local capacity to innovate:** The capacity of men and women farmers and their groups purposely to change their agricultural practices. (This will be strengthened through their involvement in PTD activities.)

- **Village institutional development:** Building or strengthening local institutions and organisations for taking a lead role in PTD activities.

- **Networking:** Exchanging information and developing collaboration on a common interest while leaving the autonomy of each partner intact.

- **Supportive linkages:** Direct contacts and collaboration between farmers and "outside" organisations to support PTD activities.

- **Process monitoring:** Monitoring of the process of local capacity building.

TRAINING METHODOLOGY

Many of the issues raised in this unit are conceptual and have a long time horizon. They call for study and reflection sessions in a workshop. Past experiences (and frustrations!) of participants in promoting farmer groups or organisations will often be a good entry point for discussion. The discussion section of this unit will give participants an overview of the issues.

Because of the complexity of the issues, iteration between workshop training and implementation in actual work situations is of special importance here. Direct interaction in the field with active farmer groups or other local institutions, showing how they have taken a lead role in carrying PTD activities forward, will always be strongly motivating.

The example of the PTD training itself may provide a challenging case for participants to study process monitoring: how outside facilitators try to hand over responsibilities to the people directly involved, thus strengthening their capacities.

LEARNING ACTIVITIES

Earlier learning activities provide outlines to develop learning activities for the issues covered in this unit. For example:

- Brainstorm on strategies/methods for sustaining PTD at local level: possible outline from Learning Activity 1, Unit 2.C (p. 121).

- Debate on the role of farmer organisations: possible outline from Learning Activity 2, Unit 3.C (p. 175).

- Simulating participatory process monitoring: possible outline from Learning Activity 4, Unit 3.D (pp. 187–8).

DISCUSSION

SUSTAINING THE PROCESS AND PHASING OUT

One of the principal goals of PTD activities is to strengthen local problem-solving capacities (see Units 1.C & D, pp. 41–58). In this way, men and women farmers can play their key role in developing sustainable agriculture and improving their livelihoods. The implication is that, when such capacities have been strengthened, interaction with support agencies may become less intensive and may change character. This is frequently referred to as "phasing out", although – as shown in Unit 4.A – requests for specific consultant roles often continue to come from well-established groups or villages. The fact that direct intensive support to farmers and their institutions is needed only for a relatively short period of time, as local institutions will take over many parts of the development activities, counters the claim of some critics that PTD support is relatively expensive.

Working towards farmers' or communities' independence from support organisations and staff, however, goes against the very nature of many of these organisations; continuation of assistance is their reason for being. This calls for serious attention to these issues right from the outset, even though it is the sixth and last cluster in the PTD framework. Handing over from the start, wherever possible, responsibility for activities to the people directly involved is one important part in this (Unit 2.A, pp. 89–101). Equally important is systematic planning and monitoring specifically related to the process of strengthening local capacities to solve problems. This unit presents some important aspects to be taken into account.

STRENGTHENING INDIVIDUAL CAPACITIES

Through their involvement in PTD activities in general, men and women farmers enhance their individual capacities and skills in addressing local problems, including:

- increased self-confidence and respect for their own knowledge – for example, through participation in interactive training sessions or in visits to other areas;
- increased analytical skills – for example, through involvement in problem–cause discussions, by learning about fundamental processes underlying what is observed in the field;
- increased experimental skills – for example, through interactions around the design of experiments, and recording and assessing results;
- skills in interacting and negotiating with outside organisations.

VILLAGE INSTITUTIONAL DEVELOPMENT

PTD transforms local experimentation from being relatively unorganised and individual to being a more systematically organised and "collective" process.

Supporting local institutional development is an important element of PTD programmes. Although a major emphasis in this is usually on the development of (semi-) formal organisations, there are many other institutional possibilities to ensure that PTD becomes part of village life: traditional leaders may get interested and become stimulating actors, discussions on innovation and experimentation for sustainable agriculture may become part of certain church-meetings, sharing of PTD activities may become part of regular agricultural markets or fairs. The emergence of a cadre of local farmer-extensionists and leaders (Unit 4.A, pp. 194–6) often forms the most visible part of such institutional development.

Evidence in, for example, Zimbabwe indicates that social problems, lack of collaboration and unclearness on leadership are seen by many farmers as the major bottlenecks preventing them from addressing the agricultural challenges at hand (Hagmann *et al.*, forthcoming). For this reason alone, support activities in Zimbabwe focus on raising social awareness, developing local institutions and leadership training, following the Training for Transformation approach (Hope *et al.*, 1984).

Unit 3.C (pp. 170–75) discusses the role of experimenter groups and how to encourage their development as one way of strengthening farmer experimentation. Careful consideration must be given to institutionalising these as the basis for sustaining the PTD process.

- The (semi-) informal experimenter groups may develop towards becoming more formal PTD-oriented organisations. It is, however, doubtful whether technology development on its own is a strong enough basis and binding factor for such a formal organisation – as experience in, for example, Burkina Faso and Mali would seem to indicate (Gubbels, pers. comm.).

- Experimenter groups may, therefore, take up various other activities such as joint processing and marketing, literacy training, or savings and credit.

- Another option is to link with already existing organisations, such as cooperatives or village assemblies, and encourage them to develop PTD functions. Careful assessment of how these organisations work should reveal whether they are interested in and suitable for PTD.

■ In general, the ultimate result will rarely be that of a single organisation in the village for all sorts of development activities. Different groupings may evolve around specific needs or activities. Finding the most effective organisational patterns is, in fact, a PTD process in its own right. New forms of collaboration will need to be tried out, based on local experiences and possibilities. Monitoring and evaluation will lead to adoption, adaptation or rejection.

FIGURE 4.1
A Venn diagram to analyse local institutions

Source: Theis & Grady (1991)

An important tool for communities and fieldstaff to analyse jointly local institutional development is the Venn diagram from the PRA toolbox (Theis & Grady, 1991). This shows relevant village institutions and individuals, represented by circles. The size of the circles and their relative position indicates their relationship and influence/importance for the issues at hand.

Fieldstaff should keep an eye out for emerging organisational structures within the village and may help experimenter groups to develop their potentials in co-ordinating activities, making decisions jointly and evaluating the outcomes. They need to ensure that the groups or organisations increasingly function independently of the outside facilitators and maintain relations with third parties themselves. Possibilities to link local experimenter groups and farmer-extensionists with other village institutions and organisations should be explored. A very good overview of issues in supporting the development of local organisations is given by Esman & Uphoff (1984).

LOCAL INFORMATION BASES

Farmers and farmer groups can continue activities without direct outsider support if the village has access to the relevant information. Traditionally, experiences are kept in the memories of the people, or are translated into songs, jokes or simple drama for the benefit of future generations. PTD practitioners seek opportunities to increase the amount of information available in the village and to widen its accessibility. Some possibilities are:

■ posting the results of diagnostic activities, resources, maps, transects made, results of problem discussions, in a prominent and central place in the village;

■ keeping the planning documents and main monitoring charts in a place where all villagers can refer to them;

■ maintaining files on the experiments: topics and why chosen, design (for idea sheets see Unit 2.D, p.133), as well as results obtained;

■ keeping a collection of own training materials that have been developed locally.

Resource materials from elsewhere may complement this collection and the local data may become part of a local, community-managed, library. Such libraries have already been successfully promoted by organisations working in other fields.

HORIZONTAL LINKAGES

Farmers' strength to continue innovating will also develop from collaboration with farmers and communities beyond their own village or area (see Unit 4.A, pp.192–206). Fieldstaff can help groups from different villages to begin meeting regularly and take on responsibility for continuing the PTD process in their own villages as well as for spreading it to other villages. For example, plans for the next year's experimentation can be exchanged and compared. Joint actions may be planned towards support agencies, local policymakers or traders. Such a forum may start when leaders of experimenter groups participate in *joint evaluation and planning meetings* on last year's experiments in one of the participating villages. NGOs in Indonesia have facilitated such meetings taking place every three months for more than 10 years now (Musante & Kingsley, unpublished). Such informal or semi-formal inter-village networks may invite collaborating institutions to send representatives. Roles and responsibilities can be coordinated and resources allocated for the coming season.

In Bassar, Togo (two years after the facilitator of World Neighbors left), 12 village communities were continuing to meet annually to analyse and evaluate the previous season's experiments and schedule the research agenda for the coming season. At this meeting they also choose a limited number of delegates to make the rounds of various agricultural development programmes and research stations, actively seeking new ideas and technologies, and to report back to their communities (Gubbels, 1988).

STRENGTHENING LINKAGES WITH SUPPORT ORGANISATIONS

In many PTD activities, fieldstaff play a key role in bringing farmers into contact with ideas and technologies available elsewhere, and other forms of support (e.g. research stations, schools, farmers' unions, or church missions). If such activities are to continue, it is important that farmers are able to maintain and expand these contacts without outside help.

In some situations the farmers' inability to gain access to support and information services may be the single most important constraint faced in developing their farm. Improving direct linkages between farmers (groups) and such services may then become a central activity in promoting innovation (CARE, 1994)

To strengthen farmers' and communities' vertical linkages, fieldstaff may:

- share their information on sources of technologies and other support; explain where they obtained the new ideas;
- encourage and facilitate direct physical contact through visits both ways and by bringing relevant publications into the villages;
- look for ways to institutionalise the linkages.

Clearly this is something that needs attention, not only in the later stages of PTD but right from the start. To be able to play this role properly, fieldstaff have to know the various organisations and their potential contribution to the PTD programme. This is possible through an "organisational inventory" or a "RAAKS" activity early in the programme (Unit 2.A, p. 92).

Thus, fieldstaff can help farmers to make contact with *extension and service organisations,* to solicit their aid in providing back-up support for farmer experiments. Networking early in the PTD process should facilitate this development. For example, forestry staff working in tree nurseries can assist in acquiring tree species that farmers want to try out, participate in trials for multiplication of species with characteristics that farmers prefer, and participate in the evaluation of farmer experiments.

Of particular importance is the back-up support from researchers and research organisations to the local experimentation programme. Such links will also open up possibilities for farmers to influence the research agenda of the research stations. Institutionalising this link becomes a reality if:

- farmer experimenters join annual research review meetings at research institutes, as done by FARMI in the Philippines (Balbarino, 1994, pers. comm.);
- leaders of experimenter groups become involved in the evaluation of research programmes;
- or become members of councils that set priorities for future research organised by the research stations;
- or become members of an expert panel with respect to trials on a particular problem area.

Strengthening direct linkages with existing mass communication media will also influence the PTD process positively and decrease farmers' dependency on outside facilitators. This may take the simple form of ensuring direct subscriptions by farmer experimenters or their groups to farmer journals or magazines. But much stronger involvement in the production of mass media has also shown potential (Box 4.6). In certain cases, farmer-extensionists may become local reporters or local "agents" of relevant journals.

MONITORING THE CAPACITY TO INNOVATE

Monitoring and evaluation of PTD activities not only looks at whether relevant technologies are being

BOX 4·6 "THE PEASANTS' MESSAGE"

In Tabacundo, Ecuador, the radio school service organised a weekly programme called *Mensaje Campesino* (the peasants' message), which is produced by farmers for farmers, with the help of "radio auxiliaries" – volunteers who have received a short training in how to operate a cassette recorder. The farmers recorded audio-materials of their own choice.

This simple initiative has encouraged the communities to increase exchanges amongst groups and communities. Twenty villages have formed local associations since the beginning of the programme.

Source: O'Sullivan-Ryan & Kaplan (1979)

developed (Unit 3.D, pp.176–88) but also whether capacities to innovate are being strengthened (often summarised under "process monitoring"). From the framework of Unit 3.D the questions of Who? and What? (and, to a lesser extent, How?) need attention in such process monitoring.

First of all, it is important to realise that there is no single answer to the *"Who"* question. Several parties will generally benefit from process monitoring, including the men and women farmers who are the main actors in the programme. Encouraging farmers, their groups and communities to review regularly and systematically what has been achieved in making them less dependent on outsiders may, in fact, be the single most important role of fieldstaff in sustaining the PTD process.

Beyond farmers and fieldstaff, other parties in process monitoring include managers of support organisations, policy-makers and the funding agencies. Each has its own legitimate monitoring need. It is a challenge in process monitoring to make sure that all these different needs are met in an integrated way. For example, by ensuring that farmers' monitoring results are accepted and give information to other levels, and that policy-level monitoring questions are translated into realistic farmer-based indicators.

Elements of *"What"* needs to be monitored can be found in the above discussion on how to strengthen capacities to innovate. For each aspect mentioned, locally relevant indicators can be found. Table 4.1 gives a number of examples.

TABLE 4·1
EXAMPLES OF INDICATORS FOR PROCESS MONITORING

AREA OF INTEREST	EXAMPLES OF INDICATORS
Self-confidence	■ number of people speaking up in meetings ■ ways in facing officials ■ value given to own ideas, own knowledge, culture
Analytical skills	■ frequency of use of analytical tools and techniques
Experimental skills	■ number of experiments ■ quality of experiments
Leadership	■ number of farmer-extensionists ■ functioning of farmer-extensionists
Organisational development	■ number of groups ■ participation in meetings and activities (who?) ■ conflict resolutions ■ own financial management
Horizontal linkages	■ participation in meetings with other villages ■ joint activities with other villages
Vertical linkages	■ number of visits by support agencies to the village ■ links with journals, etc ■ degree of institutionalisation of the links
Resource mobilisation	■ own contributions to activities; money and/or kind ■ amount of outside resources attracted

Process monitoring within the context of PTD programmes will use participatory approaches. The relevant parties will be encouraged and supported in looking at their own activities and role, first and foremost for their own benefit. A key role of outside evaluators in this will be to ask the appropriate questions in order to find relevant indicators: what does it mean when we talk about strengthened capacity, what in it is important for us? Only then can ways be found *"How"* to monitor, how to measure and how to collect information.

PTD ORGANISATIONS AS RESOURCE CENTRES

Development of its function as *resource centre* enables a PTD organisation to give critical support to existing PTD groups as well as to facilitate spreading to other organisations and areas. Documentation of the PTD process and results may include:

- compilation and processing of information with reference to experiments undertaken by farmers. Files per topic including concise information on why the experiment is important; the set-up of the experiment; the working hypotheses; the results of measurements and farmers' evaluations; conclusions (adaptation/further testing, rejection, wider diffusion, required supporting activities, alternative/ additional ideas to be tested); results of possible follow-up activities (Scheuermeier, 1988);

- development and compiling of resource materials coming out of local PTD activities or obtained from elsewhere. Preparation of these materials forces the staff members to reflect periodically on the field experience and the monitoring data, to study certain topics of specific relevance in greater detail, and to produce materials for publication, for training, for distribution to farmer trainers, etc. Such a system assures that staff members have to systematise the experiences gained, and keep in touch with literature and experiences elsewhere;

- production of "idea kits": these are collections of a great number of potentially relevant technological options, together forming a pool of ideas which the extension worker can use in dialogue with local farmers. Fieldworkers will use the information contained in the kit to develop materials in local languages and adapt illustrations, when necessary. The kit may consist of folders that each contain a great number of single-concept sheets (2–4 pages maximum per technology). Each sheet explains the basic principles with direct practical relevance to a specific technology, including only what is really important in order to understand and try out the technology. Many self-explanatory illustrations are included (Gonsalves, 1990).

- production of folk drama and audiovisuals, such as in the Kenya Woodfuel Development Programme in Kakamega District. Local actors and comedians were employed to stage an amateur drama incorporating local sayings and songs. The experiences gained in collaborative experimentation with respect to tree planting and management with a limited number of farmer groups were fed into a drama which is staged on market days. The drama provides a reflection on the woodfuel problem, tells about the experimentation done by the initial groups, and encourages participants to develop their own ideas as to how the situation can be improved (KWAP, 1991). In Vietnam, the idea of experimenting with new technological options in crop protection is spread through a rather conventional mass-media campaign: posters, leaflets, radio discussions. Careful monitoring is planned to reveal whether the fundamental message "try out for yourself and decide" is coming across beyond the technical message (Heong & Escalada, unpublished).

Frequently, pioneer PTD programmes make the greatest impact by stimulating other organisations to take up PTD seriously; spreading the PTD process by transferring it to other organisations, rather than by doing it yourself. In practice, this implies that existing programmes may act as training grounds for local key persons and staff of other organisations in the management and implementation of the PTD processes. Another possibility may be the organisation of field workshops for policy-makers, researchers or extension staff as an effective way to stimulate other institutions to replicate the PTD process. The inclusion of farmer experimenters in the workshop as participants, resource persons and trainers has proved to be a very effective way of reorientation (deschooling) of research and extension workers for appropriate role performance in PTD. Such workshops may also be helpful to sensitise senior officials and generate required institutional support.

MONITORING THE IMPACT ON THE AGRO-ECOLOGY

The fundamental assumption throughout this guide has been that increased farmer involvement enhances the possibilities for developing towards sustainable agriculture (Unit 1.A, pp. 30–40). Apart from assessment of the results of experiments and their impact on farmers' livelihood (Unit 3.D, pp. 176–88) and on the capacity of farmers to innovate (this unit), the impact of the experimentation on the sustainability of the agro-ecological system demands specific attention. To be meaningful, this assessment will be done (at least partly) together with the farmer experimenters.

Information may be obtained through the farmer evaluations of the results of each individual experiment. This implies including evaluation criteria concerning agro-ecological effects and sustainability as a whole (compare Bimbao *et al.*, 1995). In addition, certain indicators may be systematically monitored over a longer period of time in order to be able to detect certain trends. Possible examples include silt traps, total biomass production on selected farms, soil life indicators on selected fields, the quality of irrigation water, the occurrence of particular pests or predators. Such reading of the land is sometimes known as "land literacy" (Campbell, 1994). In these latter monitoring activities, other household members, schools and schoolteachers, local social organisations, etc, can make important contributions.

Periodic monitoring of the extent of the spread of certain technologies generated through the PTD process (to whom, where, how quickly, costs involved, constraints for further spreading, required follow-up, etc) is of equal importance. Such monitoring may be implemented periodically, with the help of the leaders and trainers of the experimenter groups, applying a simple format for a community survey.

In general, participatory monitoring of the changes in sustainability of a local agro-ecological system is a relatively new, yet crucial, challenge in agricultural development efforts. Readers who are pioneering approaches in this field are greatly encouraged to document and share these wherever possible.

REFERENCES, RESOURCES AND CONTACTS

APPENDIX

Appendix / References, Resources and Contacts

REFERENCES

Adoyo, F. (1994), "The mirror technique" in: *ILEIA Newsletter*, 10 (1): 22–23.

Alders, C., Haverkort, B. & Veldhuizen, L.R. (1993), *Linking with farmers: networking for Low-External-Input and Sustainable Agriculture*, IT Publications, London, UK.

Ashby, J.A. (1990), *Evaluating technology with farmers: a handbook*, IPRA-Project, International Centre for Tropical Agriculture, Cali, Colombia.

Ashby, J.A., Garcia, T., Guerro, M. del P., Patiño, C.A., Quiros, C.A. & Roa, J.I. (forthcoming), "Supporting local farmer research committees" in: *Farmers' research in practice*, IT Publications, London, UK.

Ashby, J.A, Quiros C.A. & Rivers J.M. (1990), "Farmer participation in technology development: work with crop varieties" in: Chambers, R., Pacey A. & Thrupp, L.A. (eds), *Farmer first: farmer innovation and agricultural research*, IT Publications, London, UK.

Beingolea, O.J (1994), *Planning and conducting experimental field trials in peasant communities*, World Neighbors, Santiago, Chile.

Bentley, J.W. (1992), "Promoting farmer experiments in non-chemical pest control", paper prepared for the IIED/IDS workshop Beyond Farmer First, Institute of Development Studies, University of Sussex, UK.

Biggs, S.D. (1989), *Resource-poor farmer participation in research: a synthesis of experiences from nine NAR systems*, OFCOR Comparative Study Paper 3, ISNAR, The Hague, The Netherlands.

Bimbao, M.A., Lopez, T. & Lightfoot, C. (1995), "Learning about sustainability" in: *ILEIA Newsletter*, 11 (2): 28–30.

Bliek, J. van der & Veldhuizen, L. van (1993), *Developing tools together: the role of participation in the development of tools, equipment and techniques in Appropriate Technology Programmes*, available from ITDG, Myson House, Railway Terrace, Rugby CV21 3HT, UK.

Bokkestijn, A. & Werf, E. van der (1995), "Learn, implement and evaluate: an iterative training process", in: *ILEIA Newsletter*, 11 (1): 26–27.

Box, L. (1987), *Experimenting cultivators: a methodology for adaptive agricultural research*, Agricultural Administration (Research and Extension) Network Discussion Paper 23, ODI, London, UK.

Box, L. (1989), "Virgilio's theorem: a method for adaptive research" in: Chambers, R., Pacey A. & Thrupp, L.A. (eds), *Farmer first: farmer innovation and agricultural research*, IT Publications, London, UK.

Box, L., ed. (1990), *From common ignorance to shared knowledge: knowledge networks in the Atlantic Zone of Costa Rica*, Wageningen University Studies in Sociology, Wageningen, The Netherlands.

Braidwood, R.J. (1967), *Prehistoric men*, Scott, Foresman & Co., Glenview, Illinois, USA.

Brouwers, J. (1993), *Rural people's response to soil fertility decline: the Adja case (Benin)*, Agricultural University, Wageningen, The Netherlands.

Budelman, A. (1983), *Primary agricultural research: farmers perform field trials – experiences from the Lower Tana Basin, East Kenya*, Tropical Crops Communication 3, Department of Tropical Crop Science, Agricultural University, Wageningen, The Netherlands.

Bunch, R. (1982), *Two ears of corn: a guide to people-centered agricultural improvement*, World Neighbors, Oklahoma City, USA.

Bunch, R. (1989), "Encouraging farmers' experiments" in: Chambers, R., Pacey A. & Thrupp, L.A. (eds), *Farmer first: farmer innovation and agricultural research*, IT Publications, London, UK.

Bunch, R. (1991), *Dua tongkol jagung: pedoman pengembangan pertanian berpangkal pada rakyat*, World Neighbors, Jakarta, Indonesia.

Campbell, A. (1994), *Landcare: communities shaping the land and the future*, Allen & Unwin Pty Ltd, St Leonards, NSW 2065, Australia.

Canadian Council for International Co-operation (1991), *Two halves make a whole: balancing gender relations in development*, CCIC, Ottawa, Canada.

CARE (1994), "Innovators and linking" in: *International Agricultural Development*, 14 (5): 12–14.

Chambers, R. (1980), "The small farmer is a professional" in: *Ceres*, 13 (2): 19–23.

Chambers, R. (1992), *Rural appraisal: rapid, relaxed, and participatory*, IDS Discussion Paper 311, IDS, University of Sussex, Brighton, UK.

Chambers, R., Pacey, A. & Thrupp, L.A. (1989), *Farmer first: farmer innovation and agricultural research*, IT Publications, London, UK.

Connell, J. (1990), "Farmers experiment with a new crop" in: *ILEIA Newsletter*, 6 (2): 18–19.

Conway, G.R. (1987), "The properties of agroecosystems" in: *Agricultural Systems*, 24: 95–117.

Conway, G.R., McCracken, J.A. & Pretty, J.N. (1987), *Training notes for Agroecosystem Analysis and Rapid Rural Appraisal*, International Institute for Environment and Development, London, UK.

Crone, C.D. & St John Hunter, C. (1980), *From the field: tested participatory activities for trainers*, World Education, 1414 Sixth Avenue, New York 10019, USA.

Crouch, B. (1991), "Problem-census: farmer-centred problem identification" in: Haverkort, B., Kamp, J. van der & Waters-Bayer, A. (eds), *Joining farmer experiments: experiences in Participatory Technology Development*, IT Publications, London, UK.

Davis, C.D. (1989), *Community forestry: participatory assessment, monitoring and evaluation*, Community Forestry Note 2, FAO, Rome, Italy.

Davis, C.D. (1990), *The community's toolbox: the idea, methods and tools for participatory assessment, monitoring and evaluation in community forestry*, Community Forestry Field Manual 2, FAO, Rome, Italy.

Edney, J.J. (1979), "The nuts game: a concise common dilemma analog" in: *Environmental Psychology and Nonverbal Behaviour* (1978–79): 252–254.

Ellis (1983), *Getting the community into action*, University of West Indies.

Engel, P. (1991), "Farmers' participation and extension" in: Haverkort B., Kamp, J. van der & Waters-Bayer, A. (eds), *Joining farmers' experiments: experiences in Participatory Technology Development*, IT Publications, London, UK.

Engel, P.G.H. & Salomon, M.L. (1996), *Innovation for development*, Royal Tropical Institute, Amsterdam, The Netherlands.

Esman, J.E. & Uphoff, N.T. (1984), *Local organizations: intermediaries in rural development*, Cornell University Press, Ithaca, USA.

FAO (1990), *Farming systems development: guidelines for the conduct of a training course in farming systems development*, FAO, Rome, Italy.

Farrington, J. & Martin, A. (1988), *Farmer participation in agricultural research: a review of concepts and practices*, Agricultural Administration Unit Occasional Paper 9, ODI, London, UK.

Feldstein, H.S. & Jiggins J., eds (1994), *Tools for the field: methodologies handbook for gender analysis in agriculture*, Kumarian Press, West Hartford, USA.

Feldstein, H.S. & Poats, Susan V. (1989), *Working together: gender analysis in agriculture*, vol. 1, 2, Kumarian Press, West Hartford, USA.

Flanagan, J.C. (1954), "The critical incident technique" in: *Psychological Bulletin*, 51: 327–358.

Gianotten, V. & Rijssenbeek, W. (1994), *Peasant demands: manual for participatory analysis*, Ministry of Foreign Affairs, PO Box 20061, 2500 EB The Hague, The Netherlands.

Gips, T. (1986), "What is sustainable agriculture?" in: Allen, P. & van Dussen, D. (eds), *Global perspectives on agroecology and sustainable agricultural systems: proceedings of the 6th International Scientific Conference of the International Federation of Organic Agricultural Movements*, vol. I, Agroecology Program, University of California, Santa Cruz, USA.

Gnagi, A. (1992), *Participatory Technology Development: a methodology developed in promoting local beekeeping in Ouelessebougou, Mali*, Institute of Ethnology, Bern, Switzerland (in French).

Gonsalves, J. (1990), "PTD with special reference to agriculture and natural resource management", paper for workshop Training PTD and NGDOs, ETC-NL, Leusden, The Netherlands.

GRAAP (1987), *Pour une pédagogie de l'autopromotion*, 5th ed., GRAAP, Bobo Dioulasso, Burkina Faso.

Gubbels, P. (1988), "Peasant farmer agricultural self-development: the World Neighbors experience in West-Africa" in: *ILEIA Newsletter*, 4 (3):11–14.

Hagmann, J., Chuma, E. & Murwira, K. (forthcoming), "*Kuturaya*: participatory research, innovation and extension" in: *Farmers' research in practice*, IT Publications, London, UK.

Hahn, A. (1991), *Observing with the eyes, expression with the hands: farmers study natural resource management*, Notes taken of a workshop in Burkina Faso, AGRECOL, Langenbruck, Switzerland (in French).

Haverkort, B., Kamp, J. van der & Waters-Bayer, A., eds (1991), *Joining farmer experiments: experiences in Participatory Technology Development*, IT Publications, London, UK.

Haverkort, B. & Millar, D. (1992), "Farmers' experiments and cosmovision" in: *ILEIA Newsletter*, 8 (1): 26–27.

Heinrich, G.M. (1993), *Strengthening farmer participation through groups: experiences and lessons from Botswana*, OFCOR Discussion Paper 3, ISNAR, The Hague, The Netherlands.

Heong, K.L. & Escalada, M.M. (unpublished), "Farmer participatory research in rice pest management", IRRI/Visayas State College of Agriculture, The Philippines.

Hobley, M. (1991), "Gender, class and the use of forest resources: the case of Nepal" in: Rodda, A. (ed.), *Women and the environment*, Zed Books, London, UK.

Hope, A., Timmel S. & Hodzi, C. (1984), *Training for transformation: a handbook for community workers*, Mambo Press, Gweru, Zimbabwe.

Jiggins, J. & de Zeeuw, H. (1992), "Participatory Technology Development for sustainable agriculture" in: *Farming for the future; an introduction to Low-External-Input and Sustainable Agriculture*, Macmillan, London, UK.

Kolb, H.B. & Fry, R. (1975), "Towards an applied theory of experiential learning" in: Cooper, C.L. (ed.), *Theories of group processes*, John Wiley & Sons, London, UK.

Kuyper, J.B.H. (1988), *On-farm agroforestry trials in the Kisii District*, KWDP Working Paper 13, ETC Foundation/Beijer Institute, Leusden, The Netherlands.

KWAP (1991), *Participatory agroforestry extension: a field handbook*, Kenyan Woodfuel and Agroforestry Programme, PO Box 76378, Nairobi, Kenya.

Leesberg, J. & Valencia, E. (1992), "Playing with labour" in: *ILEIA Newsletter*, 8 (4): 18–20.

Lightfoot, C. & Ocado, F. (1988), "A Philippine case on Participatory Technology Development" in: *ILEIA Newsletter*, 4 (3).

Lightfoot, C., Singh, V.P., Paris, T., Mishra, P. & Salman, A. (1990), *Training resource book for farming systems diagnosis*, process documentation of an experimental learning exercise in farming systems diagnosis of the ICAR/IRRI Collaborative Rice Research Project, ICLARM/IRRI, Manila, The Philippines.

Macdonald, M., ed. (1994), *Gender planning in development agencies: meeting the challenge*, Oxfam, London, UK.

MacRae, R., Hill, S., Henning, J. & Bentley, A. (1990), "Policies, programmes, and regulations to support transition to sustainable agriculture in Canada" in: *American Journal of Alternative Agriculture*, 5 (2): 76–92.

Mag-uugmad (1994), *A farmer-based extension system: the Mag-uugmad experience*, Mag-uugmad Foundation Inc., PO Box 286, Cebu City, The Philippines.

McCorkle, C.M., Brandstetter, R.H. & McClure, G.D. (1988), *A case study on farmer innovation and communication in Niger*, Communication for Technology Transfer in Agriculture (CTTA) Project Paper, USAID, Washington, USA.

McCracken, J.A, Conway, G.R. & Pretty, J.N. (1988), *An introduction to Rapid Rural Appraisal for agricultural development*, IIED, London, UK.

MDF (1990), *Introduction to Objective Oriented Project Planning*, Management for Development Foundation, Ede, The Netherlands.

Merrill-Sands, D. (1986), *The technology application gap: overcoming constraints in small-farm development*, FAO Research and Technology Paper 1, FAO, Rome, Italy.

Millar, D. (1990), *Training for PTD in Northern Ghana*, paper for the Workshop on Training PTD and NGDOs, ETC-NL, Leusden, The Netherlands.

Millar, D., Anamoh, B., Haverkort, B. & Hiemstra, W. (1989), *Joining hands: report of the ACDEP workshop* on Participatory Technology Development held in Bolgatanga, ACDEP/ILEIA, Tamale, Ghana.

Momsen, J.H. (1991), *Women and development in the Third World*, Routledge, London, UK.

Musante, P. & Kingsley, M.A. (unpublished), "Activities for developing linkages and cooperative exchange among farmer organisations, NGOs, GOs and researchers: case study of an NGO-coordinated IPM project in Indonesia", World Education, Jl. Tebet dalam IV F no. 75, Jakarta Selatan 12810, Indonesia.

Noordwijk, M. van (1984), *Ecology textbook for the Sudan*, Ecologische Uitgeverij, Amsterdam, The Netherlands.

Okali, C., Sumberg, J. & Farrington, J. (1994), *Farmer participatory research: rhetoric or reality?*, IT Publications, London, UK.

Okali, C., Sumberg, J.E. & Reddy, K.C. (1992), "Unpacking the package: flexible message situations", paper prepared for the IIED/IDS workshop Beyond Farmer First, Institute of Development Studies, University of Sussex, UK.

O'Sullivan-Ryan, J. & Kaplun, M. (1979), *Communication methods to promote grassroots participation: a summary of research findings from Latin America*, UNESCO, Paris, France.

Payr, G. & Sulzer, R. (1981), *Handbook for Agricultural Extension*, GTZ, Eschborn, Germany.

PMHE (1992), *Workshop on Participatory Technology Development*, PMHE Project, PO Box 154, Kandy, Sri Lanka.

PMHE (1993), *PMHE follow-up training in Participatory Technology Development*, PMHE Project, PO Box 154, Kandy, Sri Lanka.

PMHE (1994), *PMHE 2nd follow-up training in Participatory Technology Development*, PMHE Project, PO Box 154, Kandy, Sri Lanka.

Pretty, J.N. (1990), *Rapid Catchment Analysis for extension agents*, notes on the 1990 Kericho Training Workshop for the Ministry of Agriculture, Kenya, International Institute for Environment and Development, London, UK.

Pretty, J. (1994), "Alternatives Systems for Inquiry", *IDS Bulletin*, 25 (2), IDS, University of Sussex, Brighton, UK.

Pretty, J.N., Guyt, I., Thompson, J. & Scoones, I. (1995), *Participatory Learning and Action: a trainer's guide*, IIED Participatory Methodology Series, IIED, London, UK.

Propelmas (unpublished), "Report on the Tabundung Survey", internal project document, Propelmas-GKS, Sumba, Indonesia.

Quiros, C.A., Gracia, T. & Ashby, J.A. (1990), *Farmer evaluations of technology: methodology for open-ended evaluation*, IPRA Instructional Unit 1, CIAT, Cali, Colombia.

Reijntjes, C., Haverkort, B. & Waters-Bayer, A. (1992), *Farming for the future: an introduction to Low-External-Input and Sustainable Agriculture*, Macmillan, London, UK.

Rhoades, R.E. & Bebbington, A. (1991), "Farmers as experimenters" in: Haverkort, B., Kamp, J. van der & Waters-Bayer, A. (eds), *Joining farmers' experiments: experiences in Participatory Technology Development*, IT Publications, London, UK.

Richards, P. (1986), *Coping with hunger: hazard and experiment in an African rice farming system*, Allen & Unwin, London, UK.

Rocheleau, D. E. (1988), "Gender, resource management, and the rural landscape: implications for agroforestry and farming systems research" in: Poats. S.V., Schmink, M. & Spring, A. (eds), *Gender issues in farming systems research and extension*, Westview Press, London, UK.

Ruddell, E. & Beingola, J. (1995), "Towards farmer scientists" in: *ILEIA Newsletter*, 11 (1): 16–17.

Rugh, J. (1986), *Self-evaluation: ideas for participatory evaluation of rural community development projects*, World Neighbors, Oklahoma City, USA.

Saito, K. A. & Spurling, D. (1992), *Developing agricultural extension for women farmers*, World Bank Discussion Paper no. 156, Washington DC, USA.

Salas, Scheuermeier & Gottschalk (1989), *Training for PTD: crucial issues and challenges*, LBL, Lindau, Switzerland.

Scheuermeier, U. (1988), *Approach development: a contribution to participatory development of techniques based on a practical experience in Tinau Watershed Project, Nepal*, LBL, Lindau, Switzerland.

Scheurmeier, U. & Sen (1994), *Starting-up Participatory Technology Development for animal husbandry in Andhra Pradesh*, LBL, Lindau, Switzerland.

Schmitz, H., Simoes, A. & Castellanet, C. (forthcoming), "Why do farmers experiment with animal traction?" in: *Farmers' research in practice*, IT Publications, London, UK.

Scoones, I. & Thompson, J., eds (1994), *Beyond farmer first: rural people's knowledge, agricultural research and extension practices*, IT Publications, London, UK.

Selener, D., Chenier, J. & Zelaya, R. (forthcoming), *Farmer-to-farmer extension in Latin America: lessons from the field*, IIRR Latin America, AP 17-08-8494, Quito, Ecuador.

Shields, D. & Thomas-Slayter, B.P. (1993), *Gender, class, ecological decline, and livelihood strategies: a case study of Siguijor Island, the Philippines*, an ECOGEN case study, Clark University, Worcester, MA, USA.

Simaraks, S., Khammaeng, T. & Uriyapongson, S. (1986), *Farmer-to-farmer workshops on small farmer dairy cow raising*, Farming Systems Research Project, Khon Kaen University, Thailand.

SIMAS (1995), *La canasta metodológica*, Servicio de Información Mesoamericano sobre Agricultura Sostenible, Apartado Postal A-136, Managua, Nicaragua.

Simpson (1994), "The lifeblood of agricultural change" in: *ILEIA Newsletter*, 10 (1): 16.

Stephens, A. (1988), *Participatory monitoring and evaluation: handbook for training fieldworkers*, Regional Home Economics and Social Development Programme, FAO Regional Office for Asia and the Pacific, 39 Phra Atit Road, Bangkok 10200, Thailand.

Stolzenbach, A. (1992), *Reflecting in action: the logic of farmer experimentation in Mali*, Department of Extension Education, Agricultural University, Wageningen, The Netherlands (in Dutch).

Stolzenbach, A. (1993), "Farmers' experimentation: what are we talking about?" in: *ILEIA Newsletter*, 9 (1): 28–29.

Svendsen, D. & Wijetilleke, S. (1983), *Navagama: training activities for group building, health and income generation*, OEF, Washington, USA.

Tan, J.G. (1986), "A participatory approach in developing an appropriate farming system in 8 irrigated lowland villages" in: *Selected proceedings of the Kansas State University Farming Systems Research Symposium*: 215–230.

Theis, J. & Grady, H.M. (1991), *Participatory Rapid Appraisal for community development: a training manual based on experiences in the Middle East and North Africa*, Save the Children/IIED, London, UK.

Ullrich, G. & Krappitz, U. (1985), *Participatory approaches to cooperative group events: introduction and examples of application*, German Foundation for International Development (DSE), Bonn, Germany.

Vel, J.A.C, Veldhuizen, L.R. van & Petch, B. (1989), "Beyond the PTD approach" in: *ILEIA Newsletter*, 5 (1): 10–12.

Veldhuizen, L. van (1990), *Bottom-up, more than words alone? The working strategy of the Propelmas Rural Development Project*, Propelmas-Gereja Kristen Sumba, Sumba, Indonesia (in Indonesian).

Verhagen, K. (1984), *Cooperation for survival: an analysis of an experiment in participatory research and planning with small farmers in Sri Lanka and Thailand*, Royal Tropical Institute (KIT), Amsterdam, The Netherlands.

Walecka, L., Caldwell, J. & Taylor, D. (1987), *Farming systems research and extension training units vols I, II & III and trainers' manual*, Farming Systems Support Project, University of Florida, USA.

Waters-Bayer, A. (1986), "Modernizing milk production in Nigeria: who benefits?" in: *Ceres*, 19 (5): 36–39.

Welligmann, B. (1994), *Orientation workshop on farmer participation in the agricultural research-extension-continuum*, Council for Agricultural Research Policy (CARP) Agricultural Research Management Project, 114/9 Wijerama Mawatha, Colombo 7, Sri Lanka.

Werf, E. van der (1989), *Ecological farming principles*, Agriculture, Man and Ecology, Pondicherry, India.

Werf, E. van der (1994), *Report on the LEISA and PTD training programme for SNV/NOVIB partners in Uganda*, ETC-NL, Leusden, The Netherlands.

Werf, E. van der (1996), *Learning for Low-External-Input and Sustainable Agriculture: a training guide*, ETC-NL, Leusden, The Netherlands.

Werner, D. & Bower, B. (1982), *Helping health workers learn*, The Hesperian Foundation, PO Box 1692, Palo Alto, CA 94302, USA.

Werner, J. (1993), *Participatory development of agricultural innovations: procedures and methods of on-farm research*, GTZ Publication 234, Eschborn, Germany.

Woodmansee, R.G. (1984), "Comparative nutrient cycles of natural and agricultural ecosystems: a step towards principles" in: *Agricultural ecosystems – unifying concepts: 145–156*, John Wiley & Sons, New York, USA.

World Neighbors (1995), *Gender sensitivity in PRA*, World Neighbors in Action 24 (2E), World Neighbors, Oklahoma City, USA.

RESOURCES

AUDIOVISUALS AND SIMULATION GAMES

CIAT (1988): *The IPRA method*
Video presenting the experiences with farmer participatory research of an international research centre working with peasants in the highlands of Colombia. Duration: 22 min. Price: US$50.00. Available with study guide from: the Participatory Research in Agriculture Project, CIAT, AA 6713, Cali, Colombia; fax: +57-23-647243 (also available in Spanish).

CLADES (1995): *Agroecology in Latin America*
Video giving an overview of the agro-ecological approach promoted by NGO members of the CLADES network, with examples from Chile. Duration: 57 min. Available from: Consorcio Latino Americano sobre Agroecología y Desarrollo, Casilla 97, Correo 9, Santiago, Chile; fax: +56-2-233-8918.

ETC (1992): *Africulture game*
Elaborated simulation game which enables trainees to experience the complexity of farm and household management by small farmers in Southern Africa. Duration: 1½–2 days. Price: US$400.00. Available from: ETC, PO Box 64, 3830 AB Leusden, The Netherlands; fax: +31-33-494-0791.

FTPP (1994): *Handing over the stick*
Video on local institutions and customary landcare in Tanzania. Available from: Forest, Trees & People Programme (FTPP), PO Box 7005, #-750 07 Uppsala, Sweden.

ICLARM (1991): *Pictorial modelling*
Video presenting a farmer-participatory method for modelling bioresource flows in farming systems which include elements of aquaculture. Duration: 10 min. Price: US$40.00 (cassette includes additional 24-min video on aquaculture). Available with accompanying training notes from: ICLARM, MC PO Box 1501, Makati, Metro Manila 1299, The Philippines; fax: +63-2-816-3183.

ICRISAT: *Participatory research with women farmers*
Video describing why and how the international research institute ICRISAT gives a major role to women's groups in comparing new bean varieties with indigenous ones. Duration: 22 min. Price: free of charge to organisations in low-income countries. Available from: TVE, PO Box 7, 3700 AA Zeist, The Netherlands.

IIRR: *Farmer-centered extension in the Philippines*
Slide–tape programme describing experiences with farmer-led extension in NGO-supported soil and water conservation programmes in the Philippines. Available from: International Institute for Rural Reconstruction (IIRR), Information Support Unit, Silang 4118, Cavite, The Philippines ($17.00 inc. booklet). IIRR also has a great number of other sound-slides and videos; list of audiovisuals available on request.

INSAN (1991): *Mending the roof of the world*
Video presenting sustainable agriculture development in the mountains of Nepal, based on the principles of permaculture. Duration: 22 min. Available from: Institute for Sustainable Agriculture, Baneshwore 10, GPO Box 3033, Kathmandu, Nepal.

Intercooperation (1993):
We could do what we never thought we could
Good introductory video to PRA, based on PRA training in Sri Lanka. Duration: 30 min. Price: US$20.00 (US$10.00 to organisations in low-income countries). Available from: Intercooperation, Self-Help Support

Programme, 92/2 DS Sananayaka Mawatha, Colombo 8, Sri Lanka.

KIOF (1992): *Organic farming*
Video on organic farming in Kenya, including detailed directions for application of various technologies in the field. Duration: 29 min. Available from: Kenya Institute of Organic Farming (KIOF), PO Box 34972, Nairobi, Kenya.

Studio Audio Visual Puskat (1991): *Farmers' laboratory*
Video presenting a clear case for stronger farmer–scientist interaction as well as the approach taken by the Integrated Pest Management programme in Indonesia to arrive at this. Duration: 17 min. Price: US$20.00 (?). Available from: Studio Audio Visual Puskat, PO Box 75, Yogyakarta 55002, Indonesia.

Terres et Vie (1993): *Et si on écoutait la terre?*
Video reporting experiences with creative forms of farmer–scientist interaction in West Africa. Duration: 27 min. Available from: Terres et Vie, 13 Rue Laurent Delvaux, B-1400 Nivelles, Belgium.

World Neighbors (1991):
Community-based experimentation and extension
Slide series describing systematically the farmer-led extension and research approach of an NGO in Mali, contrasted with the locally common top-down extension approch. Duration: approx. 15 min. Available from: World Neighbors, 5116 North Portland Avenue, Oklahoma City, Oklahoma, 73112, USA. World Neighbors also offers other filmstrips, videos and training materials on PTD and sustainable agriculture-related topics. List available on request.

SELECTED SOURCES ON PARTICIPATORY TRAINING

Crone, C.D. & St John Hunter, C. (1980), *From the field: tested participatory activities for trainers*, World Education, USA.

Hope, A., Timmel, S. & Hodzi, C. (1984), *Training for transformation: a handbook for community workers*, vols 1–3, Mambo Press, Gweru, Zimbabwe.

Narayan, D. & Srinivasan, L. (1994), *Participatory development tool kit: training materials for agencies and communities*, World Bank, Washington, USA.

Pretty, J.N., Guyt, I., Thompson, J. & Scoones, I. (1995), *Participatory learning and action: a trainer's guide*, IIED Participatory Methodology Series, IIED, London, UK.

PRIA (1987), *Training of trainers: a manual for participatory training methodology in development*, Society for Participatory Research in Asia, 45 Sainik Farm, Khanpur, New Delhi 110 062, India.

Srinivasan, L. (1990), *Tools for community participation: a manual for training trainers in participatory techniques*, PROWESS/UNDP Technical Series, New York, USA.

Theis, J. & Grady, H.M. (1991), *Participatory Rapid Appraisal for community development: a training manual based on experiences in the Middle East and North Africa*, Save the Children/IIED, London, UK.

Ullrich, G. & Krappitz, U. (1985), *Participatory approaches to cooperative group events: introduction and examples of application*, German Foundation for International Development (DSE), Bonn, Germany.

UNICEF (1993), *Visualisation in participatory programmes (VIPP): a manual for facilitators and trainers involved in participatory group events*, UNICEF, Dhaka, Bangladesh.

CONTACTS

NETWORKS AND CONTACTS FOR INFORMATION AND SUPPORT ON PTD

Agricultural University of Wageningen, Department of Communication & Innovation Studies, Hollandseweg 1, 6706 KN Wageningen, The Netherlands (fax: +31-317-404791). Contact: Niels Roling/Annemarie Groot. Implements and supports research on all aspects of farmer learning and experimentation. Is a resource centre on the RAAKS approach. Coordinates an international MSc course with major attention to PTD. Links with universities in the South to support local capacity building.

Centre for Research and Information Exchange on Low-External-Input and Sustainable Agriculture (ILEIA), PO Box 64, 3830 AB Leusden, The Netherlands (fax: +31-33-495-1779). Contact: Peter Laban. Produces a quarterly newsletter publishing numerous articles on PTD in agriculture. Publishes books and readers on PTD. Encourages and supports research on participatory approaches to the development of sustainable agriculture.

FAO Community Forestry Unit, Viale delle Terme di Caracalla, I-00100 Rome, Italy (fax: +39-6-5225-5514). Contact: Michelle Gauthier. Works with the FTPP (see below) in developing and disseminating participatory approaches to forestry development. Has initiated research on farmers' own experimentation, which is being implemented in a number of countries in the South.

FARM-Africa Farmers' Research Project, PO Box 5476, Addis Ababa, Ethiopia (fax: +251-1-652566). Contact: Alemaheyu Konde. Uses PRA methods to analyse research and development needs of farmers in southern Ethiopia. Encourages farmer experimentation and research. Provides training support to NGOs and government agencies in PRA and PTD.

Forest, Trees and People Programme (FTPP), PO Box 7005, S-75007 Uppsala, Sweden (fax: +46-18-673420). Contact: Daphne Thuvesson. Part of the Community Forestry Unit of FAO, supports the development and spread of participatory research and planning methods with partners in the South. Publishes numerous conceptual and working papers, manuals, case studies and videos, and the *Forests, Trees & People Newsletter* (in English, French and Spanish), which often contains articles on participatory approaches to agricultural development.

Institute of Development Studies (IDS), University of Sussex, Brighton BN1 9RE, UK (fax: +44-1273 621202). Contact: Robert Chambers. One of the leading institutes, together with IIED, in developing and disseminating RRA/PRA methods. Produces numerous publications, including bibliographies on the application of PRA in agriculture, forest management and irrigation.

Intermediate Technology Development Group (ITDG), Myson House, Railway Terrace, Rugby CV21 3HT, UK (fax: +44-1788 540270). Contact: Simon Croxton. Considerable experience in promoting and implementing PTD approaches in agricultural development. Has a special interest in the use of PTD in the development of Appropriate Technology tools, equipment and techniques. Issues the quarterly journal *Appropriate Technology*, which includes articles related to PTD.

International Institute for Environment and Development (IIED), 3 Endsleigh Street, London WC1H 0DD, UK (fax: +44-171-388 2826). Contact: John Thompson (Sustainable Agriculture Programme). Many years' experience in applying participatory research and development approaches. Facilitates PRA training and gives long-term support to government institutions and NGOs. Produces the periodical *PLA Notes*, training materials (including audiovisuals) and a range of RRA/PRA reports.

International Institute for Rural Reconstruction (IIRR), Silang, Cavite, The Philippines (fax: +63-969-9937). Contact: Julian Gonsalves. International NGO that implements projects with a strong PTD perspective in a great number of countries; provides international PTD training and consultancy support; runs a well-established resource centre. Regional offices in Africa (Nairobi), Latin America (Quito), USA (New York) and Europe (Brussels).

Landwirtschaftliche Beratungszentrale (LBL), CH-8315 Lindau, Switzerland (fax: +41-52-354 97 97). Contact: Ueli Scheuermeier. Participatory approaches are central to LBL's concept of agricultural extension. Provides PRA and PTD training and advisory services. Supports PTD programmes in Africa and Asia.

Natural Resources Institute (NRI), Central Avenue, Chatham Maritime, Kent ME4 4TB, UK (fax: +44-1634-880066/77). Contact: Barry Pound/Adrienne Martin. Supports participatory research and PTD programmes in various parts of the world. Provides consultancy services and training support. Encourages research–NGO linkages.

Overseas Development Institute (ODI), Portland House, Stag Place, London SW1E 5DP, UK (fax: +44-171-393 1699). Contact: John Farrington. Implements research to support policy development re Farmer Participatory Research. Publishes numerous publications on FPR. Runs an international agricultural extension and research network and produces various network papers.

West Africa Rural Foundation (WARF), CP 13, Dakar-Fann, Senegal (fax: +221-245755). Contact: Fadel Diame. Promotes methods of participatory technology development and natural resource management in Senegal, Gambia, Guinea Bissau, Guinea and Mali. Publishes multilingual (French, English, Portuguese) newsletter *L'Atelier*, as well as manuals, reports and audiovisual materials. Gives training courses and advises rural organisations.

World Neighbors, 4127 NW 122 Street, Oklahoma, OK 73120-8869, USA (fax: +1-405-752-9393). Contact: Jethro Petit. Aims at strengthening local capacities for community development in Asia, Africa and Latin America. Has wide experience in participatory approaches to development, including farmer-led experimentation. Publishes training materials and audiovisuals suitable for use by village-level leaders, and biannual newsletter *World Neighbors in Action* in English, French and Spanish.

PERIODICALS REGULARLY FEATURING PTD CONTRIBUTIONS

L'Atelier. Bulletin on PRA and PTD methods and experiences in West Africa, including training methods. WARF/FRAO, CP 13, Dakar-Fann, Senegal (fax: +221-245755).

Forests, Trees & People Newsletter. Quarterly bulletin on community forestry and natural resource management; frequent accounts of participatory approaches. Free of charge. FTPP, Box 7005, S-75007 Uppsala, Sweden (fax: +46-18-673420).

ILEIA Newsletter. Quarterly journal focused on Low-External-Input & Sustainable Agriculture; emphasises participatory methods of technology development. Free of charge to South. ILEIA, PO Box 64, NL-3830 AB Leusden, The Netherlands (fax: +31-33-495 1779).

Journal of Farming Systems Research and Extension. Quarterly journal publishing experiences with on-farm research and extension work; also discusses methodology and other issues of interest to farming systems practitioners, administrators and trainers. Subscription rates available on request. Journal of FSR-E, Attn: George Axinn, 313 Natural Resources Building, Michigan State University, East Lansing, Michigan 48824-1222, USA (fax: +1-517-353 8994).

ODI Agricultural Administration (Research & Extension) Network Papers. Twice-yearly newsletter plus several network papers with frequent attention to farmer participatory research. Portland House, Stag Place, London SW1E 5DP, UK (fax: +44-171-393 1699).

PLA Notes. Informal journal that enables PRA practitioners throughout the world to share their field experiences and methodological innovations. Free of charge. IIED Sustainable Agriculture Programme, 3 Endsleigh Street, London WC1H 0DD, UK (fax: +44-171-388-2826).

PTD Circular. Six-monthly newsletter providing information about recent publications and ongoing activities in Participatory Technology Development in sustainable agriculture. Free of charge. ETC, PO Box 64, 3830 AB Leusden, The Netherlands (fax: +31-33-494-0791).

INDEX

A

accountability, 195
'add-on' design, 160, 162
add-options strategy, 156
Adoyo, F., 119
'Africulture', game, 59, 67
agriculture: inputs, 71; research institutes, 10; smallholder, 158; traditional, 30–1, 36, 38, 160, 186
agro-ecology, 7, 9, 13, 162; monitoring, 213
agroforestry, 163
Alders, C., 7
analysis: farmer-led, 104; participation, 114
Ashby, J.A., 72, 75, 82, 131, 172
Association of Church Development Projects (ACDEP), 14
awareness, cultural, 4

B

Barangay Scholar Programme, 201–2
'Bean Experiment, the', 67, 72
beans, varieties, 145
Bebbington, A., 146, 151
Beingolea, O.J., 158–9
Bentley, J.W., 156
'best-bet' options, 4
biases, outsiders', 102, 104
Biggs, S.D., 43
biological processes, understanding, 156–7
biophysical conditions, 26
bioresource flow diagrams, 117
Bliek, J. van der, 5
board games, 178
Bokkestijn, A., 16
Bolivia: farmer experimentation, 159; farmers, 103; PTD programme, 180
'boma-mulching', 145
Bower, B., 63
Box, L., 146, 193
brainstorming sessions, 135–6, 176, 186, 207
Brouwers, J., 146, 170
Budelman, A., 193
Bunch, R., 4, 157

Burkina Faso, 208; women farmers, 193
'buzzing' groups, 28, 128, 204

C

calendars, 178; gender-differentiated, 119
Campbell, A., 115
Canadian Council for International Cooperation, 71
Cardono, A., 206
CARE, 131, 210
cash-crops, 68–9, 71
census reports, 91
Centre for Youth and Social Development, Orissa, India, 62, 65
Centro Internacional de Agricultura Tropical (CIAT), 48, 131, 172; bean programme, 72
Chambers, R., 4, 59, 104, 146
choice making, 139
cluster cards, 151
co-operation, history, 174
Colombia: CIAT, 131, 172; El Tigre, 206
communications: community channels, 200, 202; indigenous networks, 193; local media, 198, 206
community, organisational structure, 61
Community-Based Experimentation and Extension (CBEE), 48, 55
Connell, J., 147
Connelly, Stephen, 163
continuity, 61
Conway, G.R., 60, 115, 117
cost–benefit analysis, 180
credit institutions, 10
Crone, C.D., 27, 64, 105
cross visits, 130, 192, 196–7, 202, 205
Crouch, B., 116
culture: Andean, 60; conventions, 148; identity, 62–3, 65, 93; local, 155; 'of silence', 53; perceptions, 78
cyclic flow patterns, 30, 32, 39–40

D

data: analysis, 179; climatic, 91; collection, 150, 188; harvest, 197; processing compromises, 8; recording, 176–8, 182, 187; reliability, 92; secondary, 10; storage, 209
Davis, C.D., 176
decision-making, 132, 157, 187; decentralisation, 8
Development Education and Leadership Teams in Action (DELTA), 114
development, participatory approaches, 4
diagramming, 117–18; Venn, 65, 209
dialogic communications, 75–80
diversity, seeking, 30–2
division of labour, gender, 67–9, 172
donor organisations, 7
drama, 194, 197, 209, 212

E

ecosystems, 32–3
Ecuador, Tabacundo, 211
Edney, J.J., 34
education, local, 151
elites, local, 103, 195
empowerment, 43
Engel, P.G.H., 92, 119
entomology, 10
entry-point activities, PTD, 129
environment, degradation, 30–1
equitability, 61; and ethics, 43
Escalada, M.M., 212
Esman, J.E., 209
evaluation, continuous, 20
experimentation, 5–7, 10, 13, 18, 49, 51, 69, 148, 173; agroforestry, 163; assessment, 150, 154, 159–60, 162, 175–84, 186, 188; cropping, 158; design, 167–8; ecological, 169; farmer-based, 52, 170, 193, 196, 212; field studies, 152–3; flexibility, 161; hypotheses, 129; informal, 145–7, 208; inventory, 151; local, 156;

methods, 149; motivations, 146; options, 164; recording, 120; skills, 91; social value, 155; systematic, 50, 156–7, 165
extension services/systems, 10, 36, 41, 43, 71, 124, 130, 145, 154–7; farmer-based, 192–204, 207–8

F

farm systems, 161, 180
'farmer participation', 41, 43, 46
farmers: choice criteria, 131, 137, 177–9; differentiation, 102, 103, 148; ecological, 169; experimenter groups, 170–5, 187; expert workshop, 130; extension materials, 197; needs recognition, 105; organisation, 10; participation obstacles, 58; teaching role, 107; trainers, 204; women, 70, 193
Farrington, J., 4
feedback, 6, 152, 187
Flanagan, J.C., 121
food processing, 5, 46, 60, 155
Forestry Manpower Development Consultants, Holland, 66
Foundation for Ecology, Technology and Culture (ETC), 4, 59, 71
funding, 195, 198

G

gender: analysis, 69; interest conflicts, 186; sensitivity, 52, 67–8, 70–4, 196; stereotypes, 71
Ghana: ACDEP network, 14; Mampong Valley, 117; NGO network, 13; Talensi culture, 148
Gianotten, V., 103
Gnagi, A., 156
Gonsalves, J., 212
Gracia, T., 82
Grady, H.M., 114, 117, 120, 209
Groupe de Recherche et d'Appui pour l'Autopromotion Paysanne (GRAAP), 114, 119
Gubbels, P., 103, 172, 193, 208, 210

H

Hagmann, J., 208
Hahn, A., 118
Haverkort, B., 4, 149
Heinrich, G.M., 170
Heong, K.L., 212
high-external-input agriculture (HEIA), 30–3, 131, 138
Hobley, M., 69
Honduras: experimentation, 172; insect life-cycles, 156
Hope, A., 75, 79–80, 114, 119, 208
hypotheses: experimentation, 157; formulation, 140

I

'idea kits', 212
ILEIA, Newsletter, 55
improved-design strategy, 156
India: Department of Animal Husbandry, 13, 15; landlessness, 69
indigenous technology development, 41, 44
Indonesia: NGOs, 210; Propelmas, 90PTD, 90
informal channels, 194
'informants', 13; key, 148
information: access, 210; existing sources, 95–6; processing, 212
innovations: local, 207; monitoring, 210; perception, 154; reconstruction, 148, 150; transfer, 155
inputs, high-external, 147
Institute for Development Studies, Sussex University, 4
International Centre for Living Aquatic Resources Management, (ICLARM), 128
International Crops Research Institute for the Semi-Arid Tropics (ICRASAT), 19, 48
International Institute for Rural Reconstruction (IIRR), 192, 201
interviewing, 29, 81, 115–16, 122–4, 148, 192, 205
Investigación Participativa en Agricultura (IPRA), 48
irrigation, 91, 213; facilities, 41

J

Jiggins, J., 92, 104, 120, 130
journals, farmer, 197

K

Kaplan, M., 211
Kenya: experimentation, 145–6; Kirinyaga data, 96–7; National Council of Churches (NCCK), 27; Woodfuel Development Programme, 212
Kingsley, M.A., 210
knowledge: farmer generated, 148; indigenous, 52, 59–60; 'labelled', 12; local, 150; outsiders', 129; social basis, 145
Krappitz, U., 26

L

'land literacy', 113, 115, 213
learning, experential, 11; strategy, 156
Leesberg, J., 178
Leusden, Holland: ETC workshop, 4; trainers' workshop, 6; training workshop, 56
Lightfoot, C., 118
listening techniques, 75–7, 80, 82
livelihood systems, complexity, 59–60
low-external-input and sustainable agriculture (LEISA), 14, 16, 30–3, 36, 43, 59, 131, 138, 198; principles, 39; terminology, 38

M

MacRae, R., 33
Mag-uugmad Foundation, 193
maize, 107, 109, 146, 167; experimentation, 148, 160; streak virus, 110–11
Mali, 208; experimentation, 145, 149; millet, 157
maps, 91, 96, 113, 117, 126, 178, 197
Martin, A., 4
Mazo, C.I., 206
McCracken, J.A., 114–15

media, local, 150, 155
Merrill-Sands, D., 43
migration, male, 69
Millar, D., 14, 149
millet, 157; varieties, 145
'modernisation', 150
Momsen, J.H., 69
Musante, P., 210

N

nature, mimicking, 30, 32–3
Nederlandse Organisatie Voor Internationale Betrekk, 16
Nepal, division of labour, 68
networking, 207, 210
Nicaragua, Campesino a Campesino, 197
Nigeria: experimentation, 167; milk supply, 184–6
nitrogen deficiency, 111
non-governmental organisations (NGOs), 4–5, 10, 17, 43, 92, 101, 125, 193, 195, 200; experimentation, 146; Ghana network, 13; leaders, 90
Noordwijk, M. van, 39
North Western Province Dry Zone Participatory Development Project (NWPDZ), 19
nutrients, recycling, 31
nutrition, 91
Nuts Game, the, 30, 34–5

O

O'Sullivan-Ryan, J., 211
'Objective', game, 66
Okali, C., 156
on-farm resources, 181
Opposites, game of, 45, 175
organisational inventory, PTD, 89, 92
Orozco, L., 206

P

paired comparison, 160
Participatory Learning and Action (PLA), 113
Participatory Monitoring and Evaluation Approach (PME), 176
Participatory Rural Appraisal (PRA), 114, 173; techniques, 153; toolbox, 209
participatory situation analysis (PSA), 102, 104, 106, 113, 115, 121, 126, 128
Participatory Technology Development (PTD), 4–6, 41, 44–5, 52; aims, 150; approaches, 10, 49; explaining, 57, 99; facilitators, 106, 130–1, 135, 159–60, 171, 179–80, 187; fieldworkers, 154, 156, 176, 196–8, 211–12; focuses, 176; framework, 50–1, 55; group approach, 175; linkages, 192, 209–11; participants' awareness, 138; practitioners, 61–2, 65, 68, 75, 77, 115, 149; processes, 12, 29, 53–4, 69, 140; programmes, 129, 156, 178, 199; resource centres, 212; selection criteria, 89–90, 94; start-up, 14, 161, 212; support services, 196, 208; teams, 92, 100, 132, 139; training, 7–11, 15, 17, 19, 25, 147; types, 42; variables, 162
Payr, G., 78
perception, exercises, 83
permaculture, 38
Peru: cultural conventions, 148, experimentation, 167
'phasing out', 208
Philippines: farmer-based extension, 192; FARMI institute, 210; Mindano, 169; Rural Reconstruction Movement (PRRM), 202
Pictorial Modelling, 128
planning, self-organised, 4
power relationships, 59, 66
prejudice, professional, 53, 59
Pretty, J.N., 42, 113, 116–17, 122, 132
priorities, own, 138
problem(s): census, 28, 116, 121, 125; identification, 106; nature, 102, 105, 108–10; posing, 119–20, 127; priority, 107; solving, 207–8; -tree analysis, 112
productivity, 61
'progressive farmers', 147
Promoting Multifunctional Household Environments (PMHE), 18, 146, 153, 158, 163, 172, 182–3, 187–8
'prompting questions', 159
Propelmas, Indonesia, 90

Q

Quiros, C.A., 81–2

R

ranking, 120, 126, 131, 178; techniques, 132, 139
Rapid Appraisal of Agricultural Knowledge Systems (RAAKS), 89, 92
rapport, establishment, 93
record sheets, 178
reductionism, 148
Reijntjes, C., 4, 31, 38, 147
'replications', 158
research: comparisons, 148, 152; hypotheses, 132–3, 150
resources, allocation, 8, 10
Rhoades, R.E., 146, 151
rice, 115, 169; pest control, 121
Richards, P., 146
Rijssenbeek, W., 103
risk, spreading, 60
Rocheleau, D.E., 146
role play, 98–101, 113, 123–5
Ruddell, E., 159
Rugh, J., 176

S

Saito, K.A., 70, 74, 193
Salomom, M.L., 92
Scheuermeier, Ueli, 8, 15, 45, 57, 99, 133, 212
Schnitz, H., 132
Scoones, I., 60
secondary information, 91–2
Selener, D., 194
self-organisation, 172
self-selection, 171
Servicio de Información Mesoamericano sobre Agricultura Sostenible (SIMAS), 165

skills, increase, 208
slide tapes, 192, 201–2, 206
soil: erosion, 31; fertility, 16, 33, 118, 169, 181; life, 213; living, 30, 32; types, 180
songs, 194, 197, 209
Sperling, Louise, 79
Spurling, D., 70, 74, 193
Sri Lanka, 147; experimentation, 167; group development, 173; PMHE project, 172
St John Hunter, C., 27, 64, 105
stability, 38, 61
'stakeholder' analyses, 10
statistical analysis, 180
Stephens, A., 176, 178
stepwise testing, 156, 160–1
Stichting Nederlandse Vrijwilligers (SNV), 16
Stolzenbach, A., 145, 149, 157
storytelling, 63, 150, 194
study tours, 130
Sulzer, R., 78
sustainability, 30, 33–4, 38, 52, 150, 162, 213
Svendsen, D., 83
SWOT sessions, 8
synergy, 38
systematic reflection, 11, 13

T

Tan, J.G., 4
tape-recording, 206
'target groups', 13
teak, seed germination, 163
test plots, 157–8, 161, 165, 178
testing procedures, 157
Thailand, experimentation, 147
Theis, J., 114, 117, 120, 209
Thompson, J., 60
Togo, Bassar, 210
tractors, 108–9
Training for Transformation, 208
training: activities, 13; manuals, 16; methodology, 25, 49, 59, 67, 75, 102, 113, 129, 145, 156, 176, 192, 207; spiral model, 11; strategy, 12
transects, 118, 126
transfer-of-technology (ToT), 41–5
trials: formats, 160; special conditions, 180
trust-building, 121

U

Uganda, NGOs, 16
Ullrich, G., 26
Uphoff, N.T., 209

V

Valencia, E., 178
Valigonda clusters, 15
Vel, J.A.C., 132
Veldhuizen, L. van, 5, 93
Venn diagrams, 65, 209
Verhagen, K., 172
video-cassettes, 197
Vietnam, experimentation, 212
'Village Unknown', 100–1
villages: clusters, 194; institutional develoment, 208; organisational structure, 209
virology, 10
visualisation, 20, 113–14, 121, 128, 187; linkages, 19

W

Walecka, L., 95–6, 109–11, 124
wallsheets, 178, 187
Walsum, Edith van, 71
Walters-Bayer, A., 185
wastes, application, 32
water flows, 32
Werf, E. van der, 16, 36, 38
Werner, D., 63, 158, 160
West Africa: division of labour, 68; technology development, 48
Wijetilleke, S., 83
Wilson, Nicky, 163
women, rural, 53
Woodmansee, R.G., 32
work: experiences, 28; reproductive, 67
World Neighbors, 48, 159, 210

Z

Zaire, experimentation, 172
Zambia: experimentation, 167; maize production, 111
Zeeuw, H. de, 92, 104, 120, 130
Zimbabwe, obstacles to PTD, 208
Zwier, Joel, 195, 198